Perception of Pixelated Images

Perception of Pixelated Images

Talis Bachmann
University of Tartu,
Tallinn, Estonia

AMSTERDAM • BOSTON • HEIDELBERG • LONDON
NEW YORK • OXFORD • PARIS • SAN DIEGO
SAN FRANCISCO • SINGAPORE • SYDNEY • TOKYO

Academic Press is an imprint of Elsevier

Academic Press is an imprint of Elsevier
125, London Wall, EC2Y 5AS.
525 B Street, Suite 1800, San Diego, CA 92101-4495, USA
50 Hampshire Street, 5th Floor, Cambridge, MA 02139, USA
The Boulevard, Langford Lane, Kidlington, Oxford OX5 1GB, UK

Notices
Knowledge and best practice in this field are constantly changing. As new research and
experience broaden our understanding, changes in research methods or professional practices,
may become necessary.

Practitioners and researchers must always rely on their own experience and knowledge in
evaluating and using any information or methods described herein. In using such information
or methods they should be mindful of their own safety and the safety of others, including
parties for whom they have a professional responsibility.

To the fullest extent of the law, neither the Publisher nor the authors, contributors, or editors,
assume any liability for any injury and/or damage to persons or property as a matter of products
liability, negligence or otherwise, or from any use or operation of any methods, products,
instructions, or ideas contained in the material herein.

ISBN: 978-0-12-809311-5

British Library Cataloguing-in-Publication Data
A catalogue record for this book is available from the British Library

Library of Congress Cataloging-in-Publication Data
A catalog record for this book is available from the Library of Congress

For Information on all Academic Press publications
visit our website at http://store.elsevier.com/

Working together
to grow libraries in
developing countries

www.elsevier.com • www.bookaid.org

CONTENTS

There are two major world-views, not necessarily mutually exclusive. According to one of these the build of reality is smooth and richness and variations in the world around us and within us as well as the basic processes underlying all reality, are due to the gradients of gradual changes. Diversity is carried by increments and decrements of continuous change. Waves and oscillations are the essence; borders and discontinuities are either ambiguous, represent hidden continuities, or are conventions. According to the other stance, reality with its variations comes down to discrete units, particles, quanta; apparent smoothness and continuity are either an illusion or the result of imperfection of the observational or measurement devices. This dichotomy of attitude or of the practice of approach applies to spatial as well as temporal dimensions of reality. It covers philosophy as well as natural sciences—take the monadology of Leibniz or wave-versus-quantum dispute in physics as only two of the many well-known examples. However, for an experimental psychologist or neuroscientist who studies perception, this juxtaposition may seem a bit too radical. In most cases when research on perception is at stake it is easy to show that both approaches are meaningful. From the objective perspective, the reality of our subject matter allows both, the quantal, discrete-unit-based ways of analysis or approach, as well as the paradigms acknowledging a continuous, gradual organized reality. Receptors and neurons and action potentials can be considered as the quantal reality, easy to describe and analyze as parts of digital systems processing arrays of 1 s and 0 s. Categorical perception or semantic nodes in respective networks can also be conceived of as examples of discreteness. On the other hand, waves of brain activity recorded by EEG or MEG or smooth changes of brightness or color observed in visual psychophysics are empirical examples allowing approaches that acknowledge graduality and smoothness as attributes of subject matter.

In this book we will not dwell on theoretical debates about the continuous or quantum-like nature of the underlying research realities. We just pick up a domain where quantization offers itself robustly at its face value and in a relatively widespread practice of technology and communication—*pixelated visual images*. Nowadays, pixels are everywhere, be it your computer monitor, digital photography, or a shot of the to-be-witness in criminal cases shown on TV with the facial image pixelated. This is in order to hide the personal identity. In the latter case we have an interesting situation. The original image of a visual object—a social subject—which was initially pixelated with very high resolution is additionally pixelated at a more coarse spatial level of pixelation. The actually pixelated (ie, spatially quantized) original *objective* image was formed from a spatial array of discrete, quantal changes of luminance, but this was done with such a high spatial density (very high spatial resolution) that *subjectively* this image is experienced as an area covered or filled with a continuous, smooth change of luminance and contrast. Objective spatial resolution of luminance gradients can be much better than subjective resolution, and the discrete, objective pixel-wise changes become assimilated in subjective perception in a smooth, gradual manner. Now, by applying a special transform aided by a software program on the spatial distribution of the luminous pixels it is possible to average the luminance of these original tiny pixels within larger square-shaped areas and get an image where the—now larger—pixels can be subjectively perceived and the smoothness of contrast gradients is replaced by abrupt, discrete changes of brightness. The ever-discrete objective visual depiction of the original image (carrying image contents) can become both an indiscrete, seemingly continuous change of luminance in space where it comes to subjective perception, or a discrete, quantally changing, abrupt step-like change in luminance also represented subjectively. What matters is whether the spatial scale of pixelation, where neighboring pixels carry different levels of luminance, remains below the subjective threshold of spatial contrast discrimination or not.

As in practice there are plenty of actual or potential circumstances where pixelated images have to be or happen to be perceived, it is of practical and theoretical interest to know what the psychophysical regularities and constraints characterizing the perception of pixelated images are. What the spatial and temporal limits of perceptual discrimination of pixelated images are, what the optimal conditions for maximizing information that can be extracted when perceiving the pixelated images are, how pixelated images can be used in basic research on perception and what kind of scientific knowledge has been obtained by this method; also, how the method of pixelation can be used in increasing the capacity and versatility of communication technology without compromising the minimum necessary quality of perception by human observers—these are some of the typical questions posed in experimental research on the perception of pixelated images.

At present, information about the perception of pixelated images is scattered throughout mutually separate and sometimes even isolated publications belonging to the domains of experimental psychology, neuroscience, engineering, and even art. I decided to write this book mainly because of the wish to offer a source of information where, for the first time, most of the pertinent work is brought together in a single volume. I will review this work and provide some methodological and theoretical comments. The courage to take on this task probably comes from my own experience in this topic area where I have published some experimental studies beginning from the 1980s. Partly the motivation to write this text comes from noticing that pixelation has been often regarded simply as a means of image transform equivalent to low-pass filtering performed in the Fourier or Gaussian domain. This is a mistake that sometimes may be pardoned as it does not have detrimental consequences for interpreting experimental data, but sometimes it does. Thus, I will also focus on differentiating the effects pixelation causes in comparison with other ways of spatial filtering. I wanted to prepare this volume also because pixelated images are a common feature of the habitat of a modern individual living his or her life in the midst of rich and varied stimulation carried by contemporary communication technologies. It should be useful to understand how you perceive what surrounds you.

I see the main audience of this book consisting of graduate and postgraduate students working on the problems of visual perception, (digital) image processing, visual communication, and face perception. It must be easier for them to take one book instead of searching and looking through thousands of scattered sparse sources buried among the mass of available data that is troublesome to access. I mean psychology, neuroscience, and cognitive science students. As the approach and focus of this text is contextuated in psychology and psychophysics of visual perception and cognition, the book may be of interest to IT and communication technologies specialists—both hardware- and software-oriented—who are very well informed about the topics of pixelation from the engineering and computing perspective, but somewhat less knowledgeable about what psychologists have found about this topic. Additionally, a layman who is not afraid of some technical terms and has interest toward experimental psychology may be a welcome reader of this short book.

Several colleagues and students have been an essential support and personally rewarding coworkers in researching on psychological processes by using pixelated stimuli-images. Thank you, Neeme Kahusk, John MacDonald, Søren Andersen, Endel Põder, Iiris Tuvi (Luiga), Laura Leigh-Pemberton, Triin Eamets, Hanne-Loore Härma, Carolina Murd, Kevin Krõm, et al.! I am also really grateful to Nikki Levy, Barbara Makinster, Kiruthika Govindaraju and other Elsevier people who have been very helpful and efficient in making this book a reality. And of course, this is not just a mere traditional cliché when here and now, I express my loving feelings to my family who have always been a permanent support even though I have stolen too much precious time from them in favor of the lab and the pixels staring at me from the monitors of my computers.

Talis Bachmann

CHAPTER 1

Introduction: Visual Images and How They Are Dealt With

To set a broader context for our topic, let us use a simple tripartite framework of stages ultimately leading to the process of image perception. First, there are real objects and scenes in the environment (1). As a second stage, an image is produced (taken, captured, formed) and stored representing the characteristics of these objects or scenes (2). Third, a perceiver (human or robot) processes this image and gets some more or less adequate information about the reality represented by its corresponding image (3). To proceed from (1) to (2), some means to capture reality-information must be used. The typical example is a camera—a camera allowing projection of the image on film or a set of electronic sensors capable of feeding a certain system of encoding and storage of visual data. But scanners, radars, etc. can also be used. To proceed from (2) to (3) the captured and stored image has to be presented for the perceiver whose brain performs processing of the signals from this image and builds up cognitive representation of the objects or scenes captured and reflected in the image. Thereby, an idea of the depicted reality is formed in perception. Perception of reality as a subjective (phenomenal) image thus becomes mediated by an image of reality formed by some technical means. In our case of interest, the images are digital images somewhat impoverished by virtue of consisting only of square-shaped local areas (blocks or pixels of low resolution) with uniform intensity within each pixel. And our perceivers are not robots, but human subjects with natural brains capable of visual processing by inborn mechanisms and acquired visual skills.

However, before we can get to the central subject matter of this book—human perception of pixelated images—we need to introduce some basic terms and knowledge about images, technical aspects of image processing and evaluation as well as about how the visual system of the human brain processes visual signals.

Perception of Pixelated Images. DOI: http://dx.doi.org/10.1016/B978-0-12-809311-5.00001-5

1.1 DIGITAL IMAGES AND SAMPLING

The definition of *image* is well given by Anbarjafari (Video Lectures on Digital Image Processing, University of Tartu): "An image is a two-dimensional function f(x,y), where x and y are the spatial (plane) coordinates, and the amplitude of f at any pair of coordinates (x,y) is called the intensity of the image at that level." Here we must only add that in our less abstract context the image is a depiction of reality captured by some technical means. The *intensity* at a point of space x,y in our case refers to the intensity of visible light. The local intensities of light emitted and/or reflected from real objects or scenes change smoothly (indiscretely) as we move over the space of these objects or scenes. However, "if x,y and the amplitude values of f are finite and discrete quantities, we call the image a *digital image*. A digital image is composed of a finite number of elements called pixels, each of which has a particular location and value" (Anbarjafari, *op. cit.*). An image can *represent* some object, scene, or process from which the information about local intensities is sampled. Sampling can be either a simultaneous process where all local sampled values are measured at once or a successive, step-wise process. Basically, a digital image is a numerical representation of a two-dimensional image. For instance, if you sample information useful for estimating the brightness of a chess board and use zeros (0 s) for black squares and ones (1 s) for white squares, and if you move your sampling steps so that for each square you have two steps of sampling (starting with a black square), luminance values for one row of the chess-board squares will be marked up as 0, 0, 1, 1, 0, 0, 1, 1, 0, 0, 1, 1, 0, 0, 1, 1. Now, in order to get from a one-dimensional (1D) representation to a two-dimensional (2D) representation you should perform the same procedure on the next row of squares adjacent to the first sampled row, and so on until the whole 2D area of interest is sampled. (The next row then would be described as 1, 1, 0, 0, 1, 1, 0, 0, 1, 1, 0, 0, 1, 1, 0, 0.) This way we end up with a *matrix* of intensity values, each having a spatial coordinate x,y. Understandably, the values need not include zero and, equally understandably, we can use more levels of values for characterizing each point on an image instead of the 1-bit example above. (Thus, in the case of an 8-bit sampled representation we have 256 possible values of intensity (eg, luminance) extracted from certain spatial locations.)

It is obvious that digital sampling or digitization means that a digital image is an approximation of the real object or scene it represents. Consequently, we lose information by effecting a sampling procedure. This is unless we could have an ideal sampling device, an ideal sampler, the spatial resolution of which and the spatial progression of whose sampling operation precisely follows the object or scene it attempts to sample for information about it. An ideal sampler would produce samples equivalent to the instantaneous value of the continuous signal at the sampling points of interest. (Fig. 1.1 shows the basic principle of sampling.)

Sampling is basically the reduction of a continuous signal to a discrete signal. However, the set of discrete signals is used to get information about the real object characterized by its continuous nature. Because of this it is important to know how to characterize the real object as precisely as possible on the basis of the sampled signals, or in other words, one purports to reconstruct a continuous function from samples. For this purpose, many algorithms have been used, essentially performing an interpolation. (A well-known example is the Whittaker–Shannon interpolation formula.)

There are many sources of possible *distortion* of the sampled and reproduced image compared to the original (continuously formed) object: errors of sampling by the measuring device, effects of noise, spurious new characteristics of an image, loss of information (eg, due to quantization), slowness of sampling operation, etc. Analog-to-digital converters are typically used for sampling a real continuous signal. Haphazardly, these devices have certain physical limitations and their own sources of possible distortion and noise.

Digital images are often also called *raster images* or *bitmapped images*. Pixels of the digital image are nowadays typically stored in computer memory as a raster map—a two-dimensional array of small integers. There are many purposes, needs and ways to process information contained in the raster maps. One might like to decrease the size of the file containing this information, for example for more economical storage or faster or unconstrained communication of these files. Some other user may want to transform the original object information creatively to change the original form or view of the sampled object or

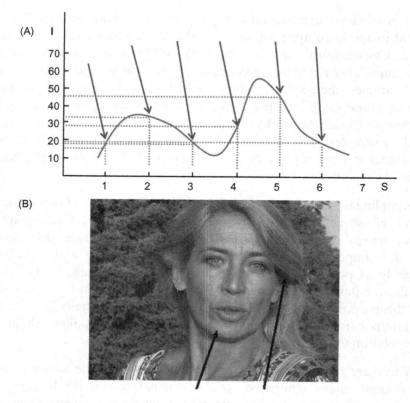

*Figure 1.1 (A) Illustration of the basics of sampling. On the ordinate we have graded levels of intensity **I** (eg, illuminance) in arbitrary units; on the abscissa we have spatial distance **S** in arbitrary units. The continuous function depicts how intensity of illuminance of some realistic object or scene changes indiscretely in space. Arrows show six steps of sampling the intensity level at spatial locations 1, 2, 3, 4, 5, and 6 yielding, respectively, **I** values equal to 16, 32, 18, 28, 46, and 19. Thus, a coarse approximation of the spatial function of luminance is obtained by picking up discrete values of luminance over certain spatial intervals. A lot of precise information is lost; for example, the highest illuminance values between steps 4 and 5 (including the peak value), the lowest values between steps 3 and 4, and also the precise dynamics of acceleration and deceleration of the function. However, if the spatial step of sampling would have been, for example, 0.001 instead of unit 1, for a naked eye the plotting of the corresponding intensity values would produce a function with a smooth, continuous appearance. Actually, the computer monitor presents this image of the apparently smooth function also in discrete values carried by each tiny pixel of the monitor light emitters. (B) An analogous operation of sampling is typically done over a 2D spatial frame. Thus, the left arrow indicates a sampling point (out of the possible large number of points) indicating a relatively higher **I** value than the **I** value sampled from the point indicated by the right arrow on a 2D image.*

scene. Somebody could find efficient methods of extraction of some critical information from the image data. Yet another person may wish to produce an image where some aspects are emphasized, etc. All these domains rely on various methods of *digital image processing* or take advantage of the developed and known practices in these. The special field of expertise in this domain focuses on algorithms of transformation that help to process digital images.

Five fundamental classes of digital image processing can be listed (Chaudhry & Singh, 2012): (1) image enhancement; (2) image restoration; (3) image analysis; (4) image compression; and (5) image synthesis. All of these are important and have met a wide variety of applications and uses, both by specialists and laypersons alike. Let me specifically comment on image compression for two reasons. First, compression is the class of image processing which is very close to our main topic—pixelated images. Second, from the practical point of view specialists as well as nonspecialists come across this topic most often. This is because by compression one obtains economy of storage space or communication load (or time) of the respective files where image information is preserved.

Image compression methods use various algorithms in order to remove data from an image. Obviously, only, or mostly, redundant data must be removed. Otherwise the compressed file will carry image information that can't be (easily) used because of its too low quality. Redundant data means that the information it carries is not relevant (for the end user) or is redundant in another sense of this term—it refers to what is known already without the stored image information. There are three main types of redundancy in digital image compression: (1) inter-pixel redundancy, (2) psychovisual redundancy, and (3) coding redundancy. Any of these redundancies or several together can be eliminated or reduced for the purposes of compression. In compressing the image, the *encoder stage* is the first and the *decoder stage* is what follows. In the encoder stage the source input is first transformed (mapper is applied) to format reducing inter-pixel redundancy in the original image and, thereafter, the quantizer reduces the accuracy of this format up to the point where infidelity of the transformed format can still be tolerated. For example, these limits can be specified by psychovisual means (perception tests for identification or categorization or quality rating): image data that are redundant for human perception can be eliminated or reduced, but beginning from a certain level of impoverishment human observers begin to notice loss of quality or necessary cues in the image. Finally, in the source encoding stage a symbol encoder is used. This consists basically of the code assignment procedure so as to represent the quantizer output by assigning short code words to the most frequently occurring output values (ie, it reduces coding redundancy). A source decoder consists of a symbol decoder and an inverse mapper. By performing the two encoding operations (mapper and symbol encoder operations) in reverse order a compressed image is produced. (Fig. 1.2 illustrates compression effects.)

Figure 1.2 The source image (original scene) at the top and its two compressed versions with progressively less detailed spatial information available. For a human observer the change in visual quality can be noticed, but the general meaning and gist of the image is essentially the same.

Compression methods are effectively possible because images contain certain statistical properties allowing economical encoding and because certain finer details of the source image can be sacrificed without loss of essential or optimal information for its depiction in the produced compressed image. A reversible compression, from where the source image can be fully restored, is called *lossless compression*. For example, when lossless compressed image data are transmitted, the receiving party can restore the original by decompressing. (For an analogous example, "8×8 alternating b w" can be restored as a matrix for a chess-board image with its black and white cells. Instead of 64 symbols for illuminance values of 0 or 1 we had only 16 symbols carrying the same information about image content.) Several well-known lossless compression methods are widely used (variable-length coding, lossless predictive coding, Huffman coding, ITU-T.6, etc.). A *lossy compression* leads to the typical possibility that after decompression the image will be quite different from the original source image. Importantly, in many cases the decompressed image with a different look is still well usable for one purpose or another. On the one hand, the crucial information can be preserved without harm, even if many other bits of visual information are lost. A simple example is when instead of a large detailed color picture a smaller gray level picture of an individual is shown. The depicted person can still be recognized. Moreover, the human mind has amazing skills in filling the gaps. This applies not only to the understanding of sentences with a few words unheard but equally well to the skill of recognizing, for example, a car, even if some parts of its image are missing or covered by a solid surface.

Among the commonly used lossy compression techniques, three can be pointed out: transform coding, lossy predictive coding, and wavelet coding. When one uses lossy compression methods repeatedly over cycles of compressing and decompressing certain image files, quality may have become severely lost in the end.

There are many commonly known and widely used image file formats differing in terms of how rich the visual information they contain is and what kind of compression is involved. The bitmap image files are a widely known and used raster graphics image file format for storing bitmap digital images. It is known also as a device independent bitmap. Its important feature is that it can be used independently of the display device that is used or available. JPEG (by Joint Photographic Expert Group) allows 8-bit grayscale and 24-bit color and is lossy. PDF (portable document format) is similar in terms of color resolution and is lossless for some types of data, but lossy for other types of data (eg, as carried by JPEG). GIF (Graphics Interchange Format) provides 1−8-bit bitonal, grayscale, or color images and is lossless. TIFF (Tagged Image File Format) is rich and varied in bit-depths and is generally lossless (an exception: JPEG inclusions). Other varieties include Flashpix, ImagePac, PNG 1.2, etc.

Digital image processing allows almost endless applications in practice and in research. Among the many tasks are pattern recognition, multiscale signal analysis, visual modeling, feature extraction, classification, and image projection. There are also many specific techniques for processing and transforming digital images such as (linear) filtering, principal or independent components analyses, wavelet analysis, pixelation, edge enhancement, variable color and artistic effects, etc.

Digital images are characterized by a couple of standard measures. *Resolution* in our context refers to the ability to distinguish fine spatial detail in the image. Technically speaking, resolution refers to the capability to measure or observe the smallest object clearly so that its boundaries can be discriminated. Resolution depends first of all on spatial sampling frequency. The smaller the spatial area of each sampling location is, the better the resolution for an observer. For a good-quality image, spatial resolution should be high. This can be measured and expressed as pixels-per-inch (ppi) or dots-per-inch (dpi) measures. Quite often certain limits are set as requirements by publishers or equipment providers for presenting images for their use. For example, 300 dpi is a typical lower bound of image accepted by peer-reviewed scientific journals. Objective spatial resolution in terms of the size of the sampled pixels can be and often is much better than the spatial resolution of the human eye and the brain systems for visual discrimination. *Dimensions* of images in terms of pixels is measured and expressed as the number of horizontally and/or vertically counted pixels per image. The total dimension in 2D is specified by multiplying the number of vertically counted pixels by the number of horizontally counted pixels (eg, 1800×1200). In other words, the number of pixel columns in the image multiplied by the number of pixel rows gives the estimate of how many pixels there are in the image. By the *dynamic range* measurement one can characterize the image in terms of how many tonal differences there are between the lightest and darkest areas of the image. One bit (two alternatives) range has only two possible illuminance values (eg, black or white, or light gray or a bit darker gray). In good-quality images there can be many thousands of spatially variable levels of light intensity. However, for practical purposes only a few or about a dozen levels are sufficient. First, this guarantees economy of storage and transmission of respective image files (small file sizes!). Second, because of the limits of visual resolution of the human visual system there is no need for "overkill" and extremely subtle intensity differences will be lost to perception anyway.

1.2 IMAGE QUALITY METRICS

Inevitably, digital images are susceptible to several types of distortion caused by a variety of factors. Distorting changes can be met at different stages of image data acquisition, processing, storage, reproduction, transmission and by multiple use. This means loss of quality of the image. It is advisable to have methods of precise measurement of image quality to ascertain whether it can still be used or not. Alternatively, valid quality measurement methods are needed also for evaluating or comparing images even without them being distorted or overused. Three main purposes of image quality assessment (IQA) can be outlined (Silpa & Aruna Mastani, 2012): (1) optimization when necessary or desired quality must be obtained at a given cost; (2) comparative analysis between different alternatives; and (3) quality monitoring in real-time applications.

Image quality assessment methods (IQA techniques) can be divided into two groups—subjective and objective (Gore & Gupta, 2015). The mean opinion score technique is recommended as one of the best subjective methods, based on opinions from a number of expert observers presented with the image which has to be evaluated. A kind of "wine tasting," isn't it? The main drawback of this technique is its high cost and slowness of its organization and administration. Both for more efficient procedures and higher level of objective standardization, objective techniques of IQA are necessary.

The standard approach in quality measurement consists in assessing the level of quality degradation of a distorted image with reference to the original image. In the case of full reference metrics (FR) one has full access to the original image. (FR methods include the signal-to-noise ratio (SNR), peak signal-to-noise ratio (PSNR), and structural similarity (SSIM) metrics. Similarity is evaluated based on a variety of possible aspects such as luminance, contrast, structural attributes, and image attribute statistics.) No reference metrics (NR) are applied when the original image is unavailable. Reduced reference metrics are somewhere between FR and NR; they are often designed to predict image quality with only partial information about the reference image available.

Many methods are pixel-based. For example, the method of mean square error is defined as a calculation of the squared difference between the pixel values of the distorted and reference image. There is a substantial shortcoming of these methods as their results do not correlate well with human subjective evaluation. This is why in recent times advanced human visual system-like models have been suggested, such as SSIM, LSDBIQ, and others (Gore & Gupta, 2015; Wang, Bovik, Sheikh, & Simoncelli, 2004; Zhang, Zhang, Mou, & Zhang, 2011). Among the recent IQMs, IFC and VIF metrics (Sheikh & Bovik, 2006; Sheikh, Bovik, & De Veciana, 2005) are evaluated as relatively close to human expert evaluation results (Wajid, Mansoor, & Pedersen, 2014). For the reader who might be interested in a more detailed account of IQM domain some published sources can be recommended: Albanesi and Amadeo (2014), Chetouani, Deriche, and Beghdadi (2010), Dosselmann and Yang (2013), Nuutinen, Halonen, Leisti, and Oittinen (2010), Oudaya Coumar, Rajesh, Balaji, and Sadanandam (2013).

For a general and more systematic treatment of digital image processing I recommend several valuable handbooks and textbooks: Burger and Burge (2013), Gonzalez and Woods (2008), Pratt (2014), Solomon and Breckon (2010).

1.3 NEUROPHYSIOLOGY OF IMAGE SAMPLING AND THE VISUAL CHANNELS

Human perception produces a subjective image with the objective visual world represented in it as smooth and continuous, with no spatially differentiated pixels noticeable. Horray, our visual system therefore seems to be a "perfect match" to the real world! Apparently then, both are conceived of as continuous, indiscrete kinds. However, if we look at our visual "machinery" more carefully (and with a powerful magnifying glass!) we immediately notice that this system itself consists of discrete units, beginning with receptors on the retina and continuing with neurons at the subcortical and cortical stages of visual information processing. How is it that this system, which is built up from discrete units, is capable of presenting a continuously changing quality of the visual space in our subjective experience? On the other hand, the objective environment quite often exposes reality in its quantized, discrete form—consider the checkerboard images as examples in the earlier parts of this text or think about computer monitors actually engineered as multipixel devices. Now, how is it that the human visual system sometimes discriminates environmental quanta (checkerboard!) and sometimes does not (computer monitor as seen from a not too close distance!)? The mystery of visual signal sampling by the human brain permitting good enough perceptual spatial resolution needs to be dealt with before we can deal with our main topic, perception of pixelated images.

It may be surprising how similar the sampling of visual signals for technical digital images is to the sampling of visual signals for the human visual system. However, in addition to some communalities such as spatial coordinates, spatial arrangement of the sensor systems, and algorithms of processing in certain information-processing networks (and certain regularities in input—output relations), the natural visual system has its own characteristics and principles of function which must be separately studied. Let us see how this system is built up and how visual signals are processed by it.

There are many stages in the human visual-processing hierarchy performing operations of transfer and transformation of signals before any subjective percept will be produced. Stimulation by light originating from the environment carries information about the quality and quantity of the visible world around us. However, besides performing useful operations for veridical perception, each of these stages is also prone to loss of information and distortion of information. In what follows I will give a brief sketch of these stages, the neurophysiological channels as the means of this information transmission, basic operations of encoding and decoding, and also of the more or less generally accepted theoretical account of spatial vision.

Objects and scenes reflect and/or emit electromagnetic waves, including the waves with wavelengths for which the human eye is sensitive and thus capable of stimulating visual systems of the brain in turn. Rays of visible light pass the eye lens and are projected onto the retina. An optical image of some object is formed on the retina as a kind of 2D "picture" of spatial luminance gradients (Robson, 1980). Point light sources in the external visual field project onto the retina so that the neighboring points in 2D space are also neighbors in the spatial map of the retina. Each external spatial point has its corresponding point at the retina. Importantly, these points can have and typically actually do have different luminance, which makes it possible to communicate information about the form, shape, pattern, and contrast gradients of the affordances of the physical world and also about stratification of the objects and features in space. Stimulation by light is converted to signals that can be transmitted and processed by the nervous system, which allows information contained in the image about objects to be acquired and processed for understanding and reacting. This stimulus "picture" can be interpreted as an array of infinitesimally small areas, each emitting light towards the eye (Robson, 1980). And here comes the first change to which the external stimulation is subjected: as the optical system of the eye (eg, the lens, the shape and elasticity of the eyeball, etc.) is not ideally perfect, the light arriving from each element of the stimulus picture does not affect only the corresponding unit in the retinal image. Its trajectory may be shifted and it undergoes spatial spread. (The light entering the eye from each point in space is spread out on the retina in such a way that the intensity of the light is highest at the center and falls off with increasing distance from the focus point. Diffraction at the pupil, focus errors due to imperfections or functional limitations of the lens are among the typical causes of this spread. In specialist literature this effect is described by the point-spread function of the optical system.)

Point light sources are not distributed in space randomly. (This is except for some special cases of natural or technological random noise-like stimulation as snowflakes in snowstorms or a noisy mess on a screen of some monitor if the apparatus is broke or perturbed.) As a rule, they are arranged as more or less regular primitives and patterns—lines, curves, stars, various shapes, more complex configurations of objects, etc. They form patterns of light. Infinitesimally large numbers of point light sources allow stimulation such that either smooth or abrupt gradients of (ill)uminance and contrast are possible in visual space. Moreover, any stimulus "picture" (stimulus image with variable spatial contrast) can be regarded as a large number of superimposed patterns of light. Accepting that any picture can be understood as a sum of a large number of super-imposed patterns of light, in the most influential psychophysical tradition a stimulus image is considered as a sum of sinusoidal pattern gratings (Weisstein, 1980) (see Fig. 1.3). In order to synthesize some specific spatial contrast-based image which itself is far from looking like a sinusoidal grating, the constituent, superimposed gratings must have a specified spatial frequency, contrast level (amplitude), orientation, and spatial phase. Not only more complex half-tone pictures (eg, as a picture in Fig. 1.1B) could be synthesized from a set of sinusoidal gratings, but also a thin line with an abrupt contrast gradient relative to the empty uniform back-ground can be produced. Similarly, a square-wave grating can be decom-posed into its fundamental sinusoidal components as harmonics (see Fig. 1.3 for an illustration). (The mathematical foundations of this approach are Fourier analysis and synthesis.)

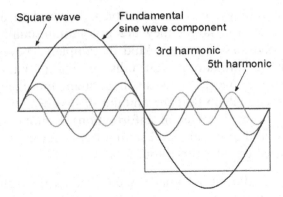

Figure 1.3 The principle of how a square-wave grating of certain fundamental frequency (here one cycle of it is shown) can be decomposed into its sinusoidal components as nth level harmonics. (Source: Courtesy of Wikipedia, Power inverter.) Bear in mind, however, that this square wave pattern shown is an ideal illustration and many necessary higher-frequency sine-wave components are not shown. In reality, the square wave periodic images synthesized from their Fourier components are just fairly good approximations of a square wave with small inhomogeneities as ripples appearing in. The more component partial sine-wave functions are used with an ever-increasing number of parameters, the more the size of these ripples will be reduced so that for practical purposes we can consider the space covered by each cycle as having homogeneous intensity.

It is necessary to acknowledge that different spatial frequency components of the stimulus light pattern (ie, how many cycles of contrast change there are per spatial unit) are not equally well represented in the image of that stimulus. Contrast of the high spatial frequency components in the image is reduced relatively more compared to the contrast of low spatial frequencies. When spatial frequency value exceeds a certain upper limit (called high cutoff frequency, with its value depending on pupil diameter and light wavelength), the image contrast becomes zero. In this case the light is spread uniformly across the image. As a consequence, very fine spatial detail cannot be communicated in the visual system. When the ratio of image contrast to the stimulus pattern contrast is calculated for the range of varying spatial frequencies we get the system's contrast *modulation transfer function*. It is common knowledge that with increasing spatial frequencies toward the higher end of the frequency values contrast of the retinal image falls off increasingly rapidly. At low spatial frequencies contrast of the retinal image is close to the contrast of the external stimulus.

When the luminous image is projected onto the retina, the light in it stimulates photosensitive receptors spread over the retina as a 2D sheet or array of units responding to light impulses. However, it would be an oversimplification to think of the retinal image as fully equivalent to the image formed from this 2D array of photoreceptors. There is a high level of homology, but not identity. Compared to light quanta, photoreceptors are too large; they are not infinitesimally small and therefore the receptors also are a kind of sampler. They sample only part of the light stimulation that falls on the retina. Fortunately enough, the size of receptors is small enough and their spacing is packed densely enough so that the amount of loss of light signals (and therefore also of environmental information) is not too dramatic. Moreover, low- and moderate-level spatial frequency contrast information is transmitted essentially without loss.

Illuminance of the retinal image is continuous through the retinal 2D space, whereas photoreceptors are discrete units, each sampling its share of the light stimulation from this continuous lighted field. Similarly to how we described image sampling in digital imaging technology, photoreceptors and higher-level units receiving signals from photoreceptors behave as sampling units producing discrete output signals. The array of such signals forms a set of point samples of a continuous function describing spatial spread of light (Robson, 1980; Williams & Hofer, 2004). However, this kind of discrete sampling is still very powerful in transferring visual information because (1) the number of photoreceptors is very high, (2) their size is very small, and (3) their response rate in time is quite high. Of course, there are many sampling artifacts, but central brain mechanisms for object recognition are very good at rejecting and reinterpreting sampling artifacts. Massive experience with visual scenes during development seems to be the "cure." Thus, despite undersampling, virtually no information is lost as long as spatial frequency components in the retinal image have cycle widths at least twice as large as the distance between adjacent receptors. In other words, the highest spatial frequency that is adequately sampled is half the sampling frequency (known as the Nyquist limit in specialist literature). Let us give some pertinent numbers. In the central part of the retina where light is projected from the point of fixation (ie, where the line of sight perpendicular to the plane of the eye lens is projected) photoreceptors are separated by about 0.002 mm. In terms of visual angle this corresponds to less than 0.5 min of arc. The area with this good sampling resolution belongs to the central foveal part of the retina. This part of the receptor mosaic can transmit retinal image information with spatial frequencies up to 60 cycles/degree of the visual angle or more. As according to the peripheral modulation transfer function spatial frequencies much above this limit cannot be transmitted anyway, the retinal receptor mosaic in the fovea does not cause any substantial loss of visual information by itself. However, the more distant from the fovea the retinal area is located, the coarser the mosaic of photoreceptors becomes. In other words, spatial resolution decreases systematically with the increase in distance from the fovea. This is *one* reason why we see fine detail of some pattern or discriminate a thin line or a small dot when directly staring at it (to bring its image at the fovea), but lose this ability when the image of these stimuli is projected on some peripheral part of the retina. Our vision is blurred. (Among other reasons why peripheral vision is blurred we find the specifics of how the ocular optical system projects images at the retina. Light from the fixation point passes directly through the center of the lens and the accommodative changes in the shape of the lens do not change the focal point of light rays from the central point source of light. Most of the light rays coming from the objects located in the periphery of the visual field do not project precisely on the surface of retina. Therefore, the corresponding images are blurred.)

The next important stage in visual signal transmission in the eye consists of the nerve cells (neurons) called ganglion cells (Deangelis & Anzai, 2004; Robson, 1980). Photoreceptor activity is transformed into bioelectrical activity and this activity is sent out of the retina by ganglion cells. (A variety of intermediate cells such as amacrine, bipolar and horizontal cells mediate processing between photoreceptors and the signals sent out from ganglion cells.) Activity of ganglion cells is expressed as nerve impulses (spikes, firing of the neuron) sent to the higher levels of the central nervous system (brain) via optic nerve fibers. Ganglion cells produce trains of impulses, typically in the range of 10−100 spikes per second (10−100 Hz frequency) on the average. Responding is brought about as a result of change in light input to the receptors connected to the ganglion cell. The stronger the stimulus, the higher the change in firing frequency. Each ganglion cell responds not only to one photoreceptor, but receives influence from many receptors. There are indirect influences and one ganglion cell can be influenced by point light stimulation from a relatively large retinal area (but not from everywhere).

The area from where the nerve cell can be effectively stimulated is termed the cell's *receptive field*. This concept can be applied to (1) spatial area in the external visual field from where stimulation can have an effect on the cell's activity, (2) the corresponding retinal area from where this cell can be influenced. Part of the ganglion cells respond by firing increase when light is switched off (or its intensity is noticeably decreased). For example, this can be obtained when a lighted area covering the center of the receptive field is suddenly made darker (eg, gray or black). These cells are classified as off-center cells. The other part of ganglion cells is responsive to light being switched on in the center of its receptive field. Respectively, these cells are referred to as on-center cells. Ganglion cells display an interesting and adaptively useful property: while stimulation of the cell's central receptive field increases the cell's activity by exciting it, stimulation close to the edge of the receptive field inhibits the cell. Thus, the system is organized according to the principle—excitatory center, inhibitory surround. (There are different subdivisions of ganglion cells depending on their spatiotemporal response profile. The so-called sustained cells or X-cells are slower and spatially more fine-tuned while the Y-cells or transient cells have larger spatial receptive fields and their temporal response is more abrupt.)

Ganglion cells and higher-level neural units connected to them (eg, the lateral geniculate body receiving input from the retina and sending input higher to cortical visual areas) are especially sensitive to contrast of the visual stimulation. Just the *difference* between luminance of the neighboring stimulation areas is what drives the neurons. As in the adaptive environment the signals carrying information about possible danger or possible positive reward are very much associated with borders of objects on the background or with patterns of texture, just the capacity of contrast discrimination proves to be adaptively important. Contrast should be rapidly and reliably detected and discriminated. The working sensitivity of the visual system neurons is reflected in how sensitive subjective visual perception is to the spatial contrast of stimulation. Sensitivity in turn can be measured and evaluated by measuring visual thresholds in psychophysical experiments. An experimenter can gradually change the objective contrast of the stimulus and measure out the contrast values specifying the lowest value still discriminable for an observer, depending on stimulation parameters. For present-day knowledge, it has been shown many times, by and large, that subjective perceptual sensitivity dependent on stimulation contrast-related variables can be sufficiently well predicted by the neuronal responses to the same stimulation (Chalupa & Werner, 2004).

Neurons of the visual system are not equally sensitive to the contrast of stimulation across different spatial frequencies (fineness of detail) of the spatial images. According to the spatial modulation transfer function where contrast sensitivity is plotted across varying spatial frequency we can say that the human eye is relatively more sensitive to contrast with spatial frequencies in the mid range (eg, about $0.1-1.0$ cycles of contrast per 1 degree of visual angle). Lower, and especially higher spatial frequency information is processed with a definitely worse sensitivity. For example, the threshold contrast value of the grating with 0.7 cycles/degree is much lower than that of the 10.0 cycles/degree grating.

Receptive fields of ganglion cells and neurons of the lateral geniculate nucleus have mostly circular spatial organizations. These cells respond to point light sources and small circular light-defined areas with on-center and off-surround. However, cortical neurons have receptive fields different in terms of their shape compared to the subcortical-level neurons. Their excitatory and inhibitory regions form parallel stripes and thus these neural units are tuned to respond to lines and more or less straight edges. Moreover, a typical cell in the area V1 in the occipital visual cortex may respond to an edge with one definite orientation (slant), but not to edges with other orientations (eg, responding to the edge line tilted 40 degrees from the 0-degree vertical, but not to lines with orientation ranging from, say, 43 to 37 degrees). The cells we are talking about respond very well to appropriately oriented grating patterns and especially well when the spatial distance between the stripes of the grating matches the distance between the excitatory or inhibitory regions of the oblongated receptive field. From V1 the flow of visual sensory signals is sent higher up to V2, V3, V4, V5/MT, etc.

Many cells in the occipital cortex respond not only to a very narrowly specified simple receptive field stimulation, but respond also to more varied types of input. For example, a stimulus capable of igniting this kind of neuron may have orientations varying over many values and/or spatial frequency content varying over coarser or finer spatial scale. Instead of responding to a static input, some cells (eg, in V5) respond only to moving stimulation. Some cortical areas are rich in cells sensitive to color and surface attributes (eg, in V4). In the temporal cortex receiving input from occipital cortex there are areas with neurons selectively responsive to face image (eg, in fusiform face area). Correspondingly, nerve cells responsive to different stimulus attributes characteristic to systematically more complex receptive fields are termed simple, complex, and hypercomplex cells. Collectively, these cells occupying different hierarchical levels of the brain compute and represent an "image" of the visual stimulation. But it is important to remember that this "image" is by no means a unidimensional representation of light intensity. Instead, we can interpret it as a set of overlapping, combined images where each region in the visual field is represented multiply by different levels tuned to different characteristics of the environmental source of the image(s) (Chalupa & Werner, 2004; Robson, 1980).

There are some general functional properties of spatiotemporal receptive fields of the primary visual system neurons (Deangelis & Anzai, 2004). Linear receptive field maps characteristic of simple cells are associated with a linear mode of transformation between visual input and neuronal response. This mode covers characteristics like orientation and spatial frequency tuning, binocular disparity tuning, temporal frequency tuning. Nonlinear response modes characteristic to nonlinear receptive field maps are proper to neurons whose responses cannot be described as linear summation of visual input across space and time. Most of this type of neurons belong to the class of complex nerve cells. Complex cells respond in the nonlinear mode of the input−output correspondences, including squaring operators. Phenomena such as surround inhibition, contextual modulation of visual input, or contrast gain control seem to be dependent on feedback connections from higher visual areas or lateral connections from within V1. The primary cortical receptive fields function as spatiotemporal image filters used for extracting certain spatiotemporal patterns from retinal images. In a sequence, a linear filter is followed by a nonlinear device and together they extract energy in the input signal that corresponds to the spatiotemporal structure of the receptive field. Linear filters are the foundation for stimulus selectivity and narrow spatial sampling. Complex neurons are freer from position or phase dependence by virtue of combining simple units across space.

In addition to point location, level of brightness, contrast, oriented lines and edges, spatial frequency, color, direction and speed of motion, binocular disparity, and depth information, some higher-level cells beyond simple and complex cells presumably can encode quite specific and at the same time very complex information at the categorical level of objects and scenes. Understandably, this can be done thanks to the orchestrated performance of the lower level, more narrowly tuned, neural cells.

Neurons of the visual system form a set of neural networks collectively performing the general function of vision. Within these networks, there are specialized channels with their receptive field characteristics tuned selectively to specific characteristics of the visual field. For our purposes, one such set of channels is especially relevant: the channels based on the arrays of neurons transmitting information about spatial frequency, orientation, or frequency bandwidth. I am talking, of course, about the concept of *spatial frequency channels* (Harris, 1980; Hess, 2004). Although based on the knowledge about spatial frequency sensitivity of the visual system neurons (including largely animal data), this concept is an abstraction used as a close approximation of what the visual system does with spatially periodic stimulation when processing it. According to this approach, the distribution of activity in processing the image of any visual scene can be calculated by convolving the object (spatial contrast) distribution with the weighing function (Robson, 1980). This is essentially a filtering operation where a hypothetical weighing function of a neuron having orientation selectivity plays the role of filter and the proximal image bears the role of data signals. As a premise, there are many different channels as filters tuned to different orientations and spatial frequencies in the visual system and there is Fourier analysis at the disposal of researchers. As a result, it became tempting to develop a theory of visual processing based on the notions of selective spatial frequency channels, spatial filtering, and Fourier domain synthesis as an explanation of how visual perceptual images of any scenes or objects are created. When we have a sufficient number of frequency- and orientation-tuned channels in the brain then the activity of the whole set can perform a Fourier transform of a retinal image; also, the activity level of each unit analyzer represents the amplitude of the component of the stimulus image having the tuned-to orientation and frequency. Thereafter, by Fourier-like synthesis, the whole picture is reconstructed for wholistic perception by combining the contributions of all selective channels. For a succinct review of the essence of Fourier analysis and synthesis as applied to visual image processing the paper by Naomi Weisstein (1980) can be recommended.

Essentially, this view suggests that visual information is analyzed by the brain in terms of the amplitude of different Fourier components of the image. As spatial frequency varies over a wide range of values, we can speak about *different spatial scales* of visual representation. There are two main ways to interpret the spatial frequency component-based view of perception (Hess, 2004). According to one view, the visual system processes information independently at different spatial scales. This principle has its advantages: when one scale is affected, it would be maladaptive in some situations if other scales also become affected and thus distort perception and/or add extra noise. Consequently, it is useful if scales are independently or differently affected. For instance, with peripheral viewing, reduced luminance, motion, stereo information, or color perception, common information derived from multiple scales will be less corrupted and perception more stable (Hess, 2004). However, if scales are independent it would be difficult or impossible to use information based on a combination of information from different scales. Furthermore, with information carried independently within different scales, the problem of scale selection may emerge. The perceiver has to decide what scale to select and what to ignore or inhibit. As we will see later in the main part of the book, with coarse quantized images, in order to recognize a face coarse scale is useful, but fine scale is detrimental for recognition. Importantly, the rules of scale selection depend on the task of perception. Among the specific "heuristics" we meet: "select the scale with best signal-to-noise ratio," "select the scale with smallest variability in filter output," or "select the scale which is typical to the category of objects."

From recent research a "rule of thumb" seems to emerge: early in the cortical visual system scale-independent processes tend to dominate, while at the level of extrastriate cortex higher up in the hierarchy scale combination becomes possible (Hess, 2004). For example, scale-invariant configural characteristics of objects are processed by the high-level nodes with large receptive fields and independently from noise processing.

Tracing back in time to Hermann von Helmholtz and reflecting the current resurgence of the Bayesian approach we must stress that visual perception is essentially a problem of *probabilistic inference* (eg, Olshausen, 2004). Properties of the environment have to be inferred from data coming from receptors. First, there is the bottom-up process of perceptual image formation. This process has to overcome uncertainty due to the fact that 3D objects and scenes have to be interpreted from the 2D spread of data on the retina and in doing this the system is perturbed with varying amounts of noise. Second, the prior probability of the state of the environment is taken into account by the processing system; perceptual and conceptual knowledge acquired by learning and former experience predicts which properties of the environment are more probable. This knowledge biases and molds perception in its turn. The visual processing system, when facing uncertainties and ambiguities of the real world, performs several types of tasks—redundancy reduction and forming sparse, overcomplete representations among the tasks. The first of these is successfully carried out by the subcortical visual mechanisms, the second by the higher-level, cortical mechanisms. Lower-level mechanisms are primarily tuned so as to transmit information about the appearance while higher-level mechanisms try to extract the meaning of the objects and scenes. If many different things or clusters of features converge on a small number of neurons, this makes a sparse code which is overcomplete and highly redundant. However, sparse coding mechanisms have also been found at the mid-level stages of visual processing. For example, some specific shapes of V1 simple-cell receptive fields can function as means for sparse representation of natural images, although at the lower level of representation a much denser set of active neurons is necessary in order to "paint" these images.

At this point I finish the brief sketch of the basics of visual image sampling and processing performed by artificial and natural visual systems. I hope that the main part on the perception of pixelated images that comes now has been granted sufficient introductory groundwork.

CHAPTER 2

Domains of Perception Research Closely Related to the Topic of Pixelation

There are several domains of research in psychophysics and experimental cognitive psychology that neighbor the topic of perceiving pixelated images. Before our treatment of perception of pixelated images begins, these "neighbors" should be introduced, which will help to better understand the subsequent, main chapters of this book. In the following part of this chapter these relevant domains will be briefly discussed.

2.1 PROCESSING SPATIAL FREQUENCY INFORMATION BY HUMAN OBSERVERS

In the mainstream psychophysics and psychophysiology, sensitivity to spatial modulation of luminance contrast has been one of the central themes. Luminance values communicated from a scene or object to an observer vary in space. Sometimes these changes are fast alternating, with different luminance values signaled from a very small spatial area. In other instances, change of the luminance value across space is slower. Within one unit of spatial area (eg, a degree of the visual angle from the observer's viewpoint) there may be only a few locations of different luminance and the spatial extent over which luminance changes may be quite large. In some spatial regions the gradient of luminance change may be unidirectional, in some other regions the increase and decrease in luminance periodically alternates. Periodically changing luminance value is best characterized by how many changes there are per unit spatial extent. This is the spatial frequency (SF) of periodic (fluctuating, oscillating) luminance change which is one of the basic characteristics of spatial contrast modulation.

A straightforward way to explain and illustrate the pertinent concepts and measures capitalizes on a specific set of images called gratings (see Fig. 2.1). First, gratings are intuitively easy to understand in terms of

Perception of Pixelated Images. DOI: http://dx.doi.org/10.1016/B978-0-12-809311-5.00002-7

Figure 2.1 Examples of visual images depicting gratings. Two left-sided images show square-wave gratings with the same orientation (the left one with higher spatial frequency and the right one with somewhat lower spatial frequency). The two images on the right show sine-wave gratings with different orientation—the left one is high-frequency grating tilted to the left (with orientation approximately at 135 = 315 degrees); the right one is low-frequency grating tilted to the right (with orientation approximately 60 = 240 degrees).

their SF, cycles, contrast, luminance modulation amplitude, etc. Second, as shown by psychophysical and sensory neuroscience work, the human brain includes sensory channels each tuned to visual input with certain SF value (Chalupa & Werner, 2004). These channels act as spatial filters. Third, there are "ready-made" analytic and computationally feasible tools well suited for quantitative analysis of spatial luminance distribution expressed as periodic functions of the changing luminance value. Fourth, because of these reasons, gratings as stimuli have been massively used in psychophysical and sensory neuroscience research.

On the left of Fig. 2.1 there are two square-wave gratings—the left one with higher *SF* (more alternating dark and light stripes along the vertical dimension) and the right one with somewhat lower SF (fewer alternating stripes). If we plot the luminance value (as a function or y-axis value) of the grating along the spatial points of its vertical dimension (as an argument- or x-axis value) we get a rectangular or *square-wave grating* spatial luminance function. Equal-size isoluminant spatial regions having low luminance level (black stripes) periodically alternate with isoluminant equal-size spatial regions having higher luminance level (white stripes). Each pair of alternating stripes consisting of one dark and one light mutually adjacent stripe represents one *cycle* of the periodic contrast modulation function. SF is measured as the number of cycles per unit spatial extent; in common practice this is defined as cycles per degree (cycles/degree) of the visual angle measured from the observer's point of view. The larger the luminance difference between the spatial areas, the higher the function value difference and concomitantly also the value of *luminance contrast*. Thus, alternating light gray and darker gray stripes would have a lower contrast level, but adjacent black and white stripes have a high contrast value. Notice that—within certain limits—an area with high overall (space averaged) luminance value may consist of periodic luminance variation with low-level contrast and an area with low overall luminance value may include luminance variation with high contrast. What matters is difference in the luminance levels. (Quite often the so-called Michelson contrast is used for contrast ratio measurement, expressed as [L(max)−L(min)]/[L(max) + L(min)].) Scaling contrast against the overall level of luminance is advisable also because of the Weber−Fechner psychophysical law showing that the absolute value of the differential threshold increases with increase in the magnitude of stimulation.

Real scenes or object images most often consist of a complex, nested combination of different spatial gradients of contrast—various high, intermediate, and low spatial frequencies are the ingredients of a compound real image. According to the Fourier analysis, every image with spatially modulated contrast can be decomposed into multiple sinusoidal (sine-wave) functions. When spatial ranges, frequencies, phases of cycles, and amplitudes of these component functions are mutually and suitably adjusted, any gray-level image can be synthesized from the underlying sinusoidal functions. Thus, even a square-wave grating can be synthesized from multiple sine-wave gratings. Two images on the right of Fig. 2.1 illustrate sine-wave gratings where luminance level changes smoothly and periodically in space. The rightmost image has relatively low spatial frequency (LSF) (SF about 1.5 cycles/degree) and its neighbor has relatively higher spatial frequency (SF about 6 cycles/degree) when observed from a typical reading distance. (Remember that the SF of gratings is measured along the axis orthogonal to the stripes that make up the grating.)

An important characteristic of the spatial contrast distribution is the shape of the spatial region involving luminance gradients. The area with variable hightened or lowered contrast may be blob-like, elongated, bar-like, stripe-like, etc. All real shapes and configurations are a combination of the local spatial shapes of varying contrast. Therefore, periodic modulation of spatial contrast carries the basic information about the environmental objects and scenes. How can you draw an object? An easy way is to put a brush on a surface and begin to move it, occasionally and/or continuously changing the direction of movement on the 2D plane of the surface. Depending on the directions and extent of change, any optional 2D shape can be produced—triangle, ellipse, square, hexagon, mollusc-like figure, outline of a car, etc. If you are allowed to take the brush off the surface and put it down at some new spatial location, the set of possible figures becomes almost limitless. The bottom line of this is that *spatial orientation* (slant) of the lines and edges of the surfaces is another basic and highly important characteristic in describing and analyzing visual images. When image processing is studied and modeled by periodic luminance contrast distribution like with gratings (or sole cycles or half-cycles of gratings), spatial orientation is naturally one of the key attributes. Orientation of the stripes and gratings composed of stripes is measured according to how big their slant is with regard to the vertical (ie, to what extent they match the vertical orientation which is taken as 0 degree). In Fig. 2.1 the left one of the sine-wave gratings has an orientation value approximately equal to 135 degrees, the rightmost sine-wave grating has an orientation of about 60 degrees, and the square-wave gratings have 90-degree orientation. Here, we must stress that in the visual system the basic afferent channels of the brain are selectively tuned both to certain SF bands of the visual stimulation and a certain relatively narrow range of spatial orientations (Chalupa & Werner, 2004). Basically, any optional visual image can be represented by a set of frequency- and orientation-tuned selective channels or filters of the visual system that collectively can process most of the image contents.

Sensitivity of the visual system is not equal for the images carrying periodic contrast modulation with different SFs. Very low frequencies (eg, below about 0.5 cycles/degree) and very high frequencies (eg, above about 20 cycles/degree) have higher luminance contrast threshold value and disappear from perception sooner when contrast is gradually reduced compared to the mid-range SFs, for which luminance contrast thresholds are lower and therefore they remain perceived when higher and lower frequencies already begin to fade out of the visual experience. For many categories of visual images (and especially for natural scenes), the higher the SF characterizing some of its content, the lower the total luminous energy contained in the contrast-modulated image for this content is. For example, for natural images, spectral power varies approximately as $1/f$ as a function of spatial frequency f (eg, Field, 1999; van der Schaaf & van Hateren, 1996). Therefore, to have equal luminous energies of LSF and high spatial frequency (HSF) contrast-modulated images or different frequency-band components of the same image, the amplitudes of the Fourier domain component functions of contrast modulation have been sometimes adjusted in the stimuli images according to the specific interest of researchers or taken into account in theorizing (eg, Bex, Solomon, & Dakin, 2009; Joubert, Rousselet, Fabre-Thorpe, & Fize, 2009).

For a more extended treatment of this approach and how it is related to psychophysics of sensation and perception, some published sources are advisable (eg, Coren, Ward, & Enns, 1999; Werner & Chalupa, 2013; Westheimer, 2012). Importantly, we have to reiterate that visual images of any real object or scene, including features and configurations of features, can be decomposed into their constituent, basic SF components. The SF, phase, and orientation of the components tell where the luminance contrast gradients are located in space; the amplitude of components tells how large the magnitude or value of luminance contrast is between the neighboring areas; SF composition tells how many cycles of fluctuating contrast change are included within a certain spatial extent, which in turn says whether the object or scene is predominantly fine-grained, coarsely robust, or uniform in terms of the predominant spatial scale, or has a more varied nature, being a mixture of components at multiple scale levels. Obviously, in order to be meaningful from the point of view of object or scene image description, these characteristics must be measured in at least 2D.

An image of a scene capturing a multitude of bushes with many branches, leaves, and textures visible in it has a high share of HSFs carrying minute detail and fine features (be they dots or thin lines, etc.). An image of a few big dark egg-shaped objects visible through a dense mist and against a uniform white background of misty snow has a high share of LSF contrast modulation (provided that viewing distance is quite close). An image of a human face observed from a close distance has a rich combination of high-frequency, mid-frequency and low-frequency components, representing, respectively, eyelashes and wrinkles (high range), eyes and mouth (mid-range), and whole head (low range). A printed text observed when reading a book is carried by high frequencies at the letter-element level. The text of the whole page, when smeared and blurred and made indiscriminable, is carried by LSFs at the page level. Thus, the predominant or defining band of SFs is a means to communicate the key information about objects and scenes. In very many cases information about the natural habitat is communicated or represented by broadband images containing many different SFs altogether. When variable SF components are equally or closely comparably represented in an image, it is the case of a "scale-invariant" expression of power spectra so that image contrast is approximately constant over widely variable frequency bands (and can be expressed by the $1/f$ type of function, where f = spatial frequency). It has been argued that in the human visual system different frequency-tuned afferent channels respond to natural scene images with equal activity (Brady, 1997). The visual system appears to have independent access to information represented by different SF channels.

However, the notion of independent and parallel channels applies primarily when the extreme cases with too high and too low frequencies are not considered and when the duration of exposure to images is optimal. When durations vary and when the top-down, task-dependent factors are considered, frequency-based bias in processing may easily happen. Philippe Schyns, Aude Oliva, and their colleagues have been the leading force behind the view that scale diagnosticity, provided the perceptual task at hand, considerably predicts whether percepts, reports, and responses are dominated by high, low, or intermediate spatial scales of the images (Morrison & Schyns, 2001; Schyns & Oliva, 1994, 1999). An experimental technique developed for this research among other methods consists of preparing and using composite (hybrid) images of objects and/or scenes so that each image is a combination of two different images—one based on high-frequency spatial information and the other based on low-frequency spatial information. From a close-range view, in the two pictures in Fig. 2.2 the fine-scale, HSF information dominates perception, showing a female face above the evil-looking male face. From a far distance (eg, move the page about 2 m away) the faces reverse their locations in perception: the evil-looking man—now carried by LSF information—is above the coarse-scale version of the female face (Schyns & Oliva, 1999).

The examples of HSF and LSF filtered images shown in Fig. 2.2 help us to see that both, HSF as well as LSF information, can be found globally over the whole area of the image. Therefore, coarse-scale, LSF information should not be taken as synonymous with global information and fine-scale, HSF information with local information. On the other hand, there is some correlation between HSF and local information as well as between LSF and global information. Namely,

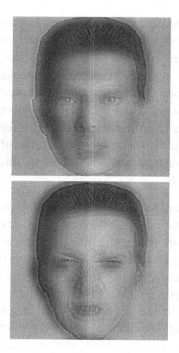

Figure 2.2 Pictures of composite images of faces. Top: composite from HSF female face and LSF evil-looking male face. Bottom: composite from HSF evil-looking male face and LSF female face. Source: Reproduced from Schyns and Oliva (1999).

very local detail of an image, such as, for example, eyelashes or edges of the nostril in a human face, cannot be represented by LSF but can be and typically are represented by HSF channels. Thus, HSF information can be both globally and locally available, but there are some limitations for LSF information to be presented from too-small a local area. In addition to the scale of coarseness (ie, SF band) and globality versus locality of depiction of image characteristics, there is also an additional, related aspect—wholistic versus component (element) distinction. In a relevant paper, Kimchi (1992) has drawn an important comparison between these aspects of image description and analysis. The terms "global" versus "local" refer to placement of visual cues within the whole area of an image (either they are depicted globally over the whole area or only from a restricted spatial locus), but the terms "wholistic" versus "component" (element) refer to *interrelations* between the image components (parts) versus the nature of individual components as such. It is easy to see that by changing the relative

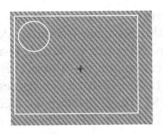 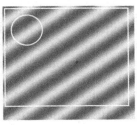

Figure 2.3 Illustration of the ways the concepts of "local versus global" and "coarse-scale versus fine-scale" are different. Global characteristics of the two similar-sized images of sine-wave gratings having different spatial frequency content (fine-scale grating on the left, coarse-scale grating on the right) are depicted within a large rectangular area compatible with the whole extent of each of the grating images. An arbitrarily chosen ring-shaped local area may include fine-scale image characteristics such as are illustrated in the high-frequency grating shown on the left or, alternatively, a same-sized ring-shaped local area may include coarse-scale image characteristics (as illustrated in the low-frequency grating shown on the right). (The local ring-shaped area within the fine-scale grating contains about 8 cycles of periodic contrast per this area along the axis orthogonal with regard to the grating stripes. This measure equals about 1 cycle for the coarse-scale grating.)

positions between the eyes, nose, and mouth the globality aspect of configuration does not change much, but the wholistic *configural* appearance of how the elements collectively look like may change considerably. (See also Piepers and Robbins, 2012.)

Sometimes, people carelessly confuse global/local and coarse-scale/fine-scale specifications of image properties. Fig. 2.3 helps to understand that this would be a mistake. The same-sized local area of an image (in this case a ring-shaped local area) may contain coarse-scale or fine-scale attributes of the image. Of course, with decreasing size of a local area there will be a certain limit to what is the coarsest SF content that can be possibly represented in that small area. For example, if we would decrease the local area within a high-frequency grating (Fig. 2.3, left) even more we still would have more than one cycle of variable local periodically changing spatial contrast, but if we decrease the same-sized local area within a low-frequency grating (Fig. 2.3, right), we could not represent the coarse-scale spatial contrast veridically any more. Globality/locality and SF (ie, fineness of detail necessary for spatial resolution) begin to interact only when the spatial window used for specifying locality is comparable in size with one cycle of the changing spatial contrast.

The relative importance of HSF and LSF information for veridical perception of objects depends also on the object categories. For example, coarse-scale information capable of communicating the basic configuration is relatively more important in fast face perception, but less effective in perceiving inanimate objects such as cars or places (Awasthi, Friedman, & Williams, 2011; Awasthi, Sowman, Friedman, & Williams, 2013; Goffaux, Gauthier, & Rossion, 2003; but see Hsiao, Hsieh, Lin, & Chang, 2005). The perception of faces compared to objects is: relatively more supported by mid-range SF bands up to 20 cycles/degree, more vulnerable to the decrease in the SF range overlap in face-matching performance, relatively more dependent on SF information than on features, and strongly dependent on coarse configural information supported by LSF (with HSF supporting more feature-based processing) (Biederman & Kalocsai, 1997; Collin, Liu, Troje, McCullen, & Chaudhuri, 2004; Goffaux, Hault, Michel, Vuong, & Rossion, 2005; Liu, Collin, Rainville, & Chaudhuri, 2000). Face recognition considerably drops when low pass–filtered images are used with SFs above 5 cycles/degree being eliminated (Fiorentini, Maffei, & Sandini, 1983). On the other hand, using more fine-scale SF information above about 8 cycles/degree does not improve face recognition compared to facial images low pass–filtered with an SF cutoff value equal to 8 cycles/degree. Visual systems seem to have highest sensitivity to facial information in the range of 8–13 cycles of contrast per facial image (Näsänen, 1999). Indeed, about 10 cycles/face is optimal, regardless of whether observers have to detect featural or configural manipulations in face images or whether eyes, nose, or mouth are manipulated (Collin, Rainville, Watier, & Boutet, 2014). While the ideal model observer performance in terms of tolerance to white noise peaked at about 5 cycles/face, human observers had consistently the best performance at 10 cycles/face. This suggests that in addition to availability of certain information in a face image, peculiarities of human perceptual processing are highly important. However, it is important to remember that the critical SF band in face recognition also depends on the size of the coarse-scale face image (Shahangian & Oruc, 2014). If the face size is less than about 2 degrees, about 3 cycles/degree is optimal for face processing. Conversely, if a blurry image of a face is large (eg, 10 degrees/face), observers cannot effectively utilize LSF information for recognition. This again reminds us that it is not SFs per se, but spatial contrast modulation per object visual size that predicts recognition and identification. For a wide range of object categories, fast and accurate identification is optimal with SFs between about 14 and 24 cycles/object (Caplette, West, Gomot, Gosselin, & Wicker, 2014). (Interestingly, more fine tuning of the parameters for best object image processing depends on the affective context of the objects.)

In general, HSF information is less important in capturing exogenous information than LSF information, is preferentially processed in the left hemisphere (especially temporal cortex), and is relatively more important for inanimate object processing, such as with places compared to faces (Awasthi et al., 2013; Carretié, Rios, Perianes, Kessel, & Álvares-Linera, 2012; Roberts et al., 2013). Different occipitotemporal regions of the brain are preferentially tuned to different SF ranges (Rotshtein, Vuilleumier, Winston, Driver, & Dolan, 2007). Interestingly, semantic information specifying a scene can influence early level interactive processing of HSF and LSF components of the image (Mu & Li, 2013). There is a prevailing view based on the results of many studies that information acquisition from images and temporal integration of different spatial scales in perception proceeds from LSF content to HSF content (Bachmann, 2000; Gao & Bentin, 2011; Goffaux et al., 2011; Kauffmann, Chauvin, Pichat, & Peyrin, 2015; Loftus & Harley, 2004; Neri, 2011; Parker, Lishman, & Hughes, 1997). However, depending on the diagnosticity of information mediated by different spatial scales for performing the perceptual task at hand, priority in processing either HSF or LSF can be flexibly changed (Sowden & Schyns, 2006).

Let us suppose indeed that coarse-to-fine spatial scale processing of visual scenes is natural for brain mechanisms. If so, then one would expect that scene-sensitive areas in the brain should show strong activation when fast sequences of filtered images of some scene will be presented, beginning with the coarsest scale image and ending with the most fine-scale image. This is in comparison with the opposite, fine-to-coarse order of presentation. Results of recent fMRI studies supported this prediction (Kauffmann et al., 2015; Musel et al., 2014). Subjects categorized these dynamic images as indoor or outdoor scenes. Coarse-to-fine presentations lead to higher activations of scene-sensitive brain areas. Particularly, parahippocampal place area, inferior frontal gyrus in the orbitofrontal cortex, fusiform area in the inferior temporal cortex, and occipital cortex in the cuneus area showed this kind of preference for coarse-to-fine order of presentation of the spectral components of a scene. Effective connectivity analysis revealed that coarse-to-fine processing indicated increased connectivity from the visual cortex to the ventral frontal area, which in turn showed increased connectivity to the inferotemporal cortex and back to the occipital cortex. These results support the processing model where initial bottom-up processing of visual signals (with transmission initially favoring coarse-scale signals in time) is followed by top-down feedback "verifying" and/or selectively emphasizing some scale-specific earlier representation. Interestingly, 7–8-month- and 12–13-month-old babies also process images according to the coarse-to-fine regularity (Otsuka, Ichikawa, Kanazawa, Yamaguchi, & Spehar, 2014). With increasing age the scale-wise temporal processing seemed to become faster and the contribution of HSFs more pronounced.

For a more thorough review of the role of SF content in facial and other types of image perception please consult reviews by Morrison and Schyns (2001), Sowden and Schyns (2006) and Ruiz-Soler and Beltran (2006).

2.2 PROCESSING GLOBAL AND LOCAL LEVELS OF FORM

Suppose you travel in the cosy English countryside and after another turn of the road a house comes to view. Seems it is an authentic Tudor-style building. What allows you to come to this conclusion? Perhaps there are various cues to your perceptual conclusion. On the one hand, the global view of the whole house may include perceptual cues allowing you to evaluate and categorize the object you see. On the other hand, certain local elements within the whole global structure such as a window or a log may also tell you something. Images of faces make a perfect example, allowing to characterize global and local levels of cues usable for recognition and categorization: the whole-face image is specifiable by its global configuration of face elements and local components or elements which also carry information that is specific to this face, but only in a restricted area—nostrils, eyes, lips, etc. Understandably, the level of locality is by no means something restricted and standard; it is even possible to zoom in or out over a hierarchy of different levels of locality within the same image. For example, an eye can be regarded as a local element within the whole-face area, but eyelashes, iris, and pupil define local elements of the eye as a relatively more global element (not to mention the whole-face frame of reference). Thus, many images are *hierarchically organized*.

Classic studies on the perception of global and local levels of visual forms were carried out by Navon (1977). He used a special class of stimuli similar to the ones shown in Fig. 2.4. Global forms such as capital letters are composed of local elements that in terms of their own

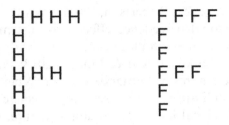

Figure 2.4 Examples of global forms (letter F) composed by arranging local forms (letters F or H) so as to form a global configuration. In one instance (right) global and local forms are congruent, in the other instance (left) a similar global form has local elements that are incongruent with the global form.

form could be either congruent with the global form or incongruent. In elegant experiments Navon showed two basic things: (1) reaction times to global forms are faster than reaction times to local forms, and (2) when perceiving a hierarchical stimulus, its global-level cues are (or begin to be) perceptually processed before the local ones. The so-called *global precedence effect* was inferred from the reaction time data. Compared to the control condition, reacting to local level form was slowed down when hierarchical stimuli were inherently incongruent (eg, large F made up from small k-s), but reacting to global-level form was not slowed down by the incongruent local forms. This asymmetry allowed Navon to conclude that global information must have been processed already when local information processing begins, but not the other way around.

The research that followed helped to specify the regularities and varieties of global/local processing in more detail. While in general the global advantage is robust over many different experimental manipulations, it is highly sensitive to attentional manipulations, alertness effects, extended practice effects, and it can be eliminated by appropriately increasing the conspicuity of the local elements, increasing the size of the global structure beyond a certain optimal limit or increasing the distance between the elements defining the global and local levels (Antes & Mann, 1984; Boer & Keuss, 1982; Deco & Heinke, 2007; Dulaney & Marks, 2007; Grice, Canham, & Boroughs, 1983; Hoffman, 1980; Kimchi, 1998; Kimchi & Palmer, 1982; Kinchla & Wolfe, 1979; Marendaz, 1985; Martin, 1979; Miller, 1981; Navon, 1981a, 1981b, 1983; Navon & Norman, 1983; Paquet, 1999; Paquet & Merikle, 1984, 1988; Rijpkema, van Aalderen, Schwarzbach, & Verstraten, 2007; Smith, 1985; Stoffer, 1993, 1994; Van Vleet,

Hoang-duc, DeGutis, & Robertson, 2011; Ward, 1982, 1983). It appears that the global precedence effect is robust mainly with short exposure durations of the stimuli and decreases with longer durations closer to 100 ms and above (Hoar & Linnell, 2013; Paquet & Merikle, 1984; however, see Andres & Fernandes, 2006). When the Navon-type of hierarchical stimuli are used for pretuning selective attention, either to the global or the local level, of stimulus structure, sensitivity to the probe gratings following the Navon stimulus is selectively increased for LSF and HSF gratings, respectively (Shulman & Wilson, 1987a). However, when the global structure is formed from local elements lacking LSFs, the global precedence effect is attenuated (Hughes, Fendrich, & Reuter-Lorenz, 1990). When Navon-type stimuli are used for priming the subsequent task of hybrid image categorization, LSF-based categorization was facilitated after priming with a global Navon task (Brand & Johnson, 2014). However, with primed facial images the Navon priming effects are controversial (Gerlach & Krumborg, 2014).

Neurobiological and ecological regularities support the principle that in the perceptual microgenesis—temporal formation of a visual percept after the stimulus image is presented—there is an initial processing advantage to low spatial resolution information and globally defined targets (Conci, Töllner, Leszczynski, & Müller, 2011; De Cesarei & Loftus, 2011; Hughes, Nozawa, & Kitterle, 1996; Kauffmann et al., 2015; Musel et al., 2014; however, see Hübner, 1997). The processing bias among the global and local characteristics as based on the spatial scale of images is determined not in terms of the absolute values of the SF content, but in terms of their relative values (Flevaris, Bentin, & Robertson, 2011a, 2011b).

Electrophysiological signatures sensitive to local processing may be extended in time beyond the signatures associated with global processing only in some experimental conditions (Han, Fan, Chen, & Zhuo, 1997), but not necessarily in some other conditions (Heinze & Münte, 1993). It appears that the global processing dominance originates at the perceptual-attentional level rather than at the response competition level (Riddernikhof & van der Molen, 1995). In the brain, there is a global processing bias in the right hemisphere and local bias in the left hemisphere (Fink et al., 1996; Gable, Poole, & Cook, 2013; Volberg & Hübner, 2004). This kind of hemispheric specialization interacts with stimulus saliency and task (Bardi, Kanai, Mapelli, & Walsh, 2012; Martens &

Hübner, 2013). Interestingly, right/global bias and left/local bias both characterize also processing image attributes within objects, whether this object is presented exclusively either to the right or to the left hemisphere (Christie et al., 2012). It should be kept in mind, however, that interactions between global- and local-level processing channels producing mutual interference effects or attentional tuning effects between levels need not be based necessarily on the direct interactions between channels carrying different SF content (Lamb & Yund, 1996a, 1996b; Lamb, Yund, & Pond, 1999). Apparently, parietal and temporal cortical areas home the mechanisms responsible for tuning perception to the global or local level of the visual structure (Romei, Driver, Schyns, & Thut, 2011; Thomas, Kveraga, Huberle, Karnath, & Bar, 2012).

When the globality versus locality issue was extended from the within-object interactions to the between-objects interactions, with global levels representing scenes, the global context effect on local processing was nevertheless obtained (Antes, Penland, & Metzger, 1981).

When attentional precues are used prior to the hierarchical Navon-type stimuli, global precues covering the whole image area facilitate global as well as local processing, but local precues facilitate only processing at the local level of form (Robertson, Egly, Lamb, & Kerth, 1993). It is possible that the initial processing for global and local image characteristics goes on in parallel, but subsequently attention determines the processing priority (Boer & Keuss, 1982; Heinze & Münte, 1993; Paquet, 1999; Stoffer, 1994; Ward, 1982). Yet, pretuning to either level of image representation can also be effective independently of voluntary attention (as manipulated by level predictability), depending on implicit readiness (Lamb, London, Pond, & Whitt, 1998; Lamb, Pond, & Zahir, 2000). Masked priming shows global processing priority already at the preconscious level (Koivisto & Revonsuo, 2004). Although high perceptual conspicuity of the local elements of a structure can lead to relatively faster local processing, structural relations among the local elements acquire the leading role when conspicuity is experimentally controlled (Love, Rouder, & Wisniewski, 1999).

The list of stimulus-related and task-related variables that affect global versus local processing can be extended even further. The extent to which inter-level effects take place considerably depends on whether the local elements or the global structure of hierarchical stimuli represent meaningful, identifiable objects or not

(Beaucousin et al., 2011; Poirel, Pineau, & Mellet, 2006). Objectness matters the most. The prevalence of global processing typically characterizes processing of novel objects, whereas local processing is typical for processing the details of the familiar perceptual input (Förster, 2012). High cognitive load reduces global bias, but only when stimuli durations are unlimited, suggesting that global prevalence is characteristic of the early stages of construction of a perceptual structure, but not of the stage of its maintenance (Hoar & Linnell, 2013). There is definite flexibility in the emphasis with which perceptual content is processed at different levels of globality. Diagnosticity of the perceptual attributes with regard to performing the task can be associated either with the local or global levels and the perceptual-attentional system is flexible in using them, depending on the task demands (Large & McMullen, 2006; Miellet, Caldara, & Schyns, 2011; Morrison & Schyns, 2001). Obviously, there are also individual differences and differences between different global/local experimental paradigms in studying level-specific processing (Dale & Arnell, 2013).

2.3 PROCESSING CONFIGURATION AND COMPONENT FEATURES

It is trivial that the subjective perceptual appearance of a hierarchical object consisting of different local features changes considerably when the mutual spatial arrangement of these features is changed. Thus, configuration of elements is an important cue for perceptual identification, recognition, and categorization. This is in addition to the individual characteristics of local features that also carry information about the identity of the object. In relative terms, configural cues are more dependent on global processing than local (element) cues are. Investigating the relative roles of local features and holistic global configuration has been one of the main themes in the psychophysics and neuroscience of perception (Davies, Ellis, & Shepherd, 1981; Maurer, Le Grand, & Mondloch, 2002; Maurer et al., 2007; Sergent, 1984; Tanaka & Farah, 1993). Historically, this goes back to the seminal studies of the Gestalt school of psychology demonstrating that the holistic structure of elements tends to dominate in perception over the impact the individual elements have.

Configural cues for object perception are relational: even with invariant features a change in the metric relations among features changes perceptual identity and appearance. One class of objects where configural processing is especially important is faces. Perception of the configural cues relies more on the LSF content of an image, while a relatively higher SF content is utilized in feature processing (Goffaux et al., 2005). Similarly to the regularities of global and local processing, processing of the configural cues is related more to the right-hemisphere brain processes and feature processing to the left-hemisphere processes (Maurer et al., 2007). It appears also that configural information is better carried by photographs and other gray-level images than by line drawings (Leder, 1996). The importance of the holistic information is stressed by the face composite effect—an invariant part of the face is perceived differently when it is combined with a different other part (Schiltz & Rossion, 2006; Young, Hellawell, & Hay, 1987). However, there is no "one-way traffic" in the interaction between the component elements and the whole composed of the elements because elements also influence how configuration is perceived and may be effective on their own (Amishav & Kimchi, 2010; Bruyer, 2011). Contribution of the local featural cues to the element/whole interaction depends on local-cue discriminability (Goffaux, 2012). An important local-feature cue which remains present also with LSF information and which can contribute to image discrimination is the shape of the local feature (Gilad-Gutnick, Yovel, & Sinha, 2012). Gestalt grouping of the different parts of a face significantly contributes to the holistic processing (Curby, Goldstein, & Blacker, 2013).

When easy and fast object recognition or discrimination is at stake, the capacity of an observer to capitalize on the wholistic (holistic) configural cues is highly important. This capacity tends to develop when objects belonging to a certain category have been extensively overlearned so that an observer becomes an expert perceiver with that category (Bukach, Gauthier, & Tarr, 2006; Diamond & Carey, 1986; Maurer et al. 2002; Palmeri & Gauthier, 2004). As face is an exceptionally overlearned visual object for virtually all individuals in the human population, it is natural that wholistic configuration-based perception is especially well developed for faces. Yet, despite the efficiency and versatility of face processing by the visual system it seems that holistic processing is most pronounced and useful from a certain range of viewing distances. For example, Ross and Gauthier (2015) estimate that the distance for effective face identification as based on holistic cues is about 2–10 m.

In the brain, configural information based on the LSF content of facial images ignites neurons in the face fusiform area in the temporal cortex (Zhao et al., 2014). The occipital face area tends to be tuned to local components carried by HSF as well as to configural information supported by LSF. It is also noteworthy that spectral sensitivity to images of objects belonging to some categories (faces, cars) is orientationally anisotropic: perception is tuned more to horizontally oriented image cues (Jacques, Schiltz, & Goffaux, 2014).

We must bear in mind that configuration is only one attribute of the holistic object processing and that the effects found with different experimental paradigms pertaining to the topic of holistic processing need not be crystal clear and universal (Burton, Schweinberger, Jenkins, & Kaufmann, 2015; McKone et al., 2013; Watson & Robbins, 2014). The role of configural information, especially in familiar face recognition, may sometimes be overestimated. Moreover, the very concept of "holistic perception" and its application to various domains of object processing have not been understood in a unanimously same way (Maurer et al., 2002; Richler, Palmeri, & Gauthier, 2012). In the present text we try to follow the distinction between the concepts "configural" and "(w)holistic," where this is essential and meaningful depending on the specific context of discussion. The term "configural" helps to stress the spatial relations (geometry) of the elements or local features of an image, that is, their relative spacing without paying much attention to the nature of the local elements or features. The term "(w)holistic" can be used when the integral set of local image characteristics forming a global structure is characterized so that all cues are taken into account. In dealing with the topic of perception of pixelated images the concepts of configuration and wholistic structure can often be used interchangeably. This is because with coarse-scale and intermediate-scale pixelation the characteristics of local features are mostly dissolved within the elementary pixels and the coarse configuration of the regions of contrast gradients has become the main cue for image identification or recognition. Yet, this may not be the case in all aspects of treating the perception of pixelated images. For example, as we will see subsequently, the pixelation-transform brings in spurious, new local cues explicated as the edges and corners of the local square-shaped areas (blocks) produced as a result of pixelation.

transform. It is advisable to use version (2) because it is intuitively clearer and also more meaningful in terms of the perceptual effects. The pixels/image measure more directly gives an impression of how severe or mild the effect of the transform could be for an observer in terms of the capability of picking up and discriminating the information that was carried by the original image before its transform. Moreover, by using pixels/image measure it is easy to compare the effects imposed upon the image by pixelation approximately to the effects brought about by standard sine-wave spatial frequency filtering. By convention, and intuitively acceptably, the filtering transform producing the certain number of pixels per image approximately mimics the Fourier filtering transform equal to half of the value of the pixels/image transform (Costen, Parker, & Craw, 1994; Harmon & Julesz, 1973).

In the examples depicted in Fig. 3.2 we have an original (source-, base-) image (left), fine quantized image with pixelation level at about 26 pixels/face along the horizontal dimension, and coarse quantized image with pixelation level at about 10 pixels/face along the horizontal dimension. The 26 pixels/face conventionally corresponds to the 13 spatial frequency cycles/face and the 10 pixels/face corresponds approximately to the 5 cycles/face. (Because in image processing and identification literature *spatial frequency per 1 degree of visual angle* is an important, meaningful and staple measure, it may be necessary in addition to the pixels/image or cycles/image, to also calculate and know what the image transform means in terms of this measure. To arrive at this measure, it is simply needed to divide the value of

Figure 3.2 Examples of facial images (from left to right): original version, fine quantized version, coarse quantized version. Number of square-shaped pixels per image progressively decreases. Source: Original image courtesy of the Stirling University pool of facial images.

cycles/image by the value of how many degrees of visual angle the image subtends. Obviously, this measure comes not directly from the size of the image itself, but from the value of the size scaled by the viewing distance. Roughly speaking, from a distance from which modern desktop computer monitors are typically viewed (eg, 60 cm), 1 degree of the visual angle corresponds to about 1 cm (to be more precise, 1 cm corresponds to 1.047 degrees). Thus, for example if your object or scene image occupies 10 cm and you observe it from 60 cm, and if this image is depicted with spatial contrast resolution equal to 12 pixels/image, your image resolution must be roughly at about 0.6 visual spatial contrast cycles/degree. If you move closer, this value decreases and if you move to a greater distance, this value increases. By convention, the value of spatial contrast cycles (either pixels/image or sine-wave spatial frequency cycles/image) is more often measured along the horizontal dimension. It is more meaningful and advisable not to measure this with regard to the canvas, but with regard to the object depicted within this canvas.)

The typical effects produced by the filtering transforms aimed at blurring the source image by eliminating high spatial frequency information can be seen in Fig. 3.3. Although the original source image (the leftmost in Fig. 3.2) became less distinctive after filtering, blurring the already pixelated image (the rightmost in Fig. 3.2) somewhat restores perceiving it as a true facial image, although carried by its low spatial frequency information.

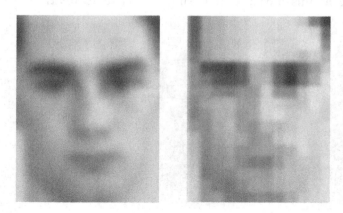

Figure 3.3 Examples of the appearance of images blurred by spatial frequency filtering with widely used Gaussian or low-pass sine-wave filters. The left image is obtained by blurring the original from Fig. 3.2, the image on the right is obtained by blurring the coarse quantized picture from Fig. 3.2.

By transforming the original image into a spatially quantized version, Harmon and Julesz (1973) aimed at producing two principal effects significant from the perception point of view. First, with a relatively coarse level of quantization (not many pixels/face defining the image) high spatial frequency information of the image is filtered out and, because fine local detail is "dissolved" within the relatively large pixels, only low spatial frequency information of the source image is communicated by the pixelated image. This is, in many respects, similar to other low-pass filtering transforms such as Fourier filtering where high spatial frequency sinusoidal components of the image spatial contrast distribution are eliminated and components whose spatial frequency is less than the cutoff value of the low-pass filter are left intact within the image. Second, in addition to the changes in the image brought about by low-pass filtering, high spatial frequency information "alien" to the information of the original image is added: the square-shaped pixels bring in new high spatial frequency information carried by the sharp edges of the squares. Essentially, by the pixelation transform a compound image is created consisting of the low spatial frequency information (coarse-scale information) of the original image and high spatial frequency information additionally carried by the mosaic of squares. This forms a second source of increased difficulty of efficient perception of the original gist of the image: high spatial frequency information different from information of the original image has a masking effect on the information preserved from the original image. By blurring (ie, low-pass filtering) of the now quantized image the masking effect is diminished (masking cues themselves are filtered out) and some recognizability of the original image is restored (Harmon, 1971; Harmon & Julesz, 1973).

However, there is also a third kind of change in addition to filtering effects and masking effects taking place between channels tuned to different spatial frequencies. This third effect is often overlooked. By creating the squares-mosaic, the configuration of the elements of the original image is also distorted. In the Fourier-filtered or otherwise blurred image many local minima and maxima of contrast present in the original image have kept their relative location within the whole image, whereas in the coarse quantized image spatial location of the contrast maxima and minima within the isoluminant squares becomes uncertain and ambiguous. Moreover, the mosaic of the relatively large local squares produces a new, competing configuration (a "blocked" appearance stretching and twisting the local gradients of luminance that were present in the original image). I will discuss this problem also in the following parts of the text.

3.2 WHAT KIND OF IMAGES ARE TYPICALLY PIXELATED AND WHY

Typically, the images subjected to pixelation transform are gray-level, half-tone pictures, such as photographs or gray-level computer images. Human faces are by far the most often used types of images pixelated in various research contexts and applied contexts. In basic research different spatial frequency components have been purposely manipulated by pixelation in order to study the effects of the different spatial levels of facial information on face identification, face recognition, emotional expression recognition, visible speech processing, etc. (Bachmann, 1991, 2007; Bachmann & Kahusk, 1997; Bachmann & Leigh-Pemberton, 2002; Bhatia, Lakshminarayanan, Samal, & Welland, 1995; Bindemann, Attard, Leach, & Johnston, 2013; Campbell & Massaro, 1997; Costen et al., 1994; Costen, Parker, & Craw, 1996; Demanet, Dhont, Notebaert, Pattyn, & Vandierendonck, 2007; Hanso, Murd, & Bachmann, 2010; Harmon, 1973; Harmon & Julesz, 1973; Lakshminarayanan, Bhatia, Samal, & Welland, 1997; Lander, Bruce, & Hill, 2001; Lu & Sperling, 1996; MacDonald, Andersen, & Bachmann, 2000; Morrone & Burr, 1997; Morrone, Burr, & Ross, 1983; Nurmoja, Eamets, Härma, & Bachmann, 2012; Parker & Costen, 1999; Sergent, 1986; Sinha, Balas, Ostrovsky, & Russell, 2006; Sperling & Hsu, 2004; Tieger & Ganz, 1979; Torralba, 2009; Uttal, Baruch, & Allen, 1997; Vitkovich & Barber, 1996; Wallbott, 1991, 1992; White & Li, 2006). Stimulus images other than faces have also been used in their pixelated form, including scenes with objects, dynamic images, and images transformed from stimuli having only two levels of luminance, such as letters (eg, Berry, Kean, Misovich, & Baron, 1991; Borgo et al., 2010; Hallum, 2007; Morrone & Burr, 1997; Nandakumar & Malik, 2009; Parker, Lishman, & Hughes, 1996a, 1996b; Sommerhalder, 2007; Torralba, 2009; Uttal, Baruch, & Allen, 1995a, 1995b, 1995c).

Apart from the purposeful application of pixelation in basic research, the topic of perception of pixelated images emerges also in a more applied context by default. Even though the modern advanced imaging technology allows high-resolution and reliably captured images of interest in principle, in practice this is not always the case. Images taken by security system cameras, surveillance systems, satellite-based cameras, occasional photography usable as forensic evidence, mobile phone installed cameras, and so on now and then lack sufficient quality. Even though the quality of equipment may be high, what is left for observation and analysis is of low quality for some other reasons. (For a related discussion see Bindemann et al., 2013; Bruce & Young, 1998; Burton, Wilson, Cowan, & Bruce, 1999; Busey & Loftus, 2007; Frowd et al., 2005; Henderson, Bruce, & Burton, 2001; Knoche, McCarthy, & Sasse, 2008; Lee, Wilkinson, Memon, & Houston, 2009; Sinha, 2002a, 2002b; Yang, Kriegman, & Ahuja, 2002; Zhao & Chellappa, 2006). Often this means that the spatial resolution of an image limits it to the relatively coarse-scale pixelated image. Even when the quality and reliability of the technology as such is very good, conditions of photographic or video recording do not permit to use this capacity up to its full potential. The distance to the object may be too big; recording may be carried out in noisy conditions, the signals may be too weak for various reasons. Again, all this may result in an image very much similar to what we observe when looking at a coarse quantized picture. Thus, to get to know the limits human perception has in extracting useful information from the quantized images and to know what kind of requirements the imaging equipment should correspond to (in order to be useable by human observers) are important practical aims. Moreover, it is useful to ascertain what kinds of behavioral/psychophysical procedures or perception conditions could optimize or maximize human performance when they have to work with pixelized images.

In real-life conditions, pixelated images can be frequently met. Most often this happens when TV broadcasts, magazines or newspapers want to or are forced to hide the personal identity of somebody depicted by their media. Protection of personal information may be necessary either because of possible infringement of human rights, for ethical reasons, because of the potential threat to the person, or due to the need to hide the identity of professionals working in law protection systems. Related to this, it is important to know what the pixelated image characteristics are that guarantee the anonymity of the depicted person. Sometimes items of art also depict pixelated images. A well-known example belongs to Salvador Dali who in 1974–76 used the famous pixelated portrait of Abraham Lincoln when he created his mosaic art depicting his wife Gala. More exercises of mosaic art followed. If such a style of depiction is used in art, people may want to know the regularities of perception associated with such images.

3.3 HOW A STUDENT OR A LAYMAN CAN GENERATE PIXELATED IMAGES

The software market has a variety of options for those interested in effecting image pixelation. Many standard image-processing packages include commands for image pixelation. For example, Adobe Photoshop and Paint Shop Pro software both have the Mosaic transform function for this purpose. Irfan View software includes an analogous Pixelize function. They are all relatively easy and straightforward to use and quick to learn. Among the many tips for how to use them for preparing stimulus material for demonstrations and experiments there are a few that are most useful. First, never loose the original image by effecting some transform on it, having forgot to keep the separate saved file of the original. This is especially important when you apply lossy transformation to the original image. Second, it is advisable to save interim products of image transformations separately. Third, if there are many steps in working with multiply transformed images and you want to be sure that you have not inadvertently changed something so you do not know what the change was or do not want to confuse images, use written records specifying the essence of the image and/or chronology of operations (either on the sheets including the images, or as file names, or both.) It is not forbidden to use handwritten notes for these purposes.

If the purpose of work presupposes very high quality of images or very good discrimination between images, high-quality computers and monitors have to be used. The quality level of the workstation or other computing device used for image processing should be no less than the facilities supporting the end-users—for example, publishers, poster production, web sources, projection devices, photography, etc. But the best recommendation is this: the more you practice with digital image processing focused on pixelation applications, the more you know the tricks, pitfalls and surprising resources of this domain.

CHAPTER 4

Unmasking Pixelated Images

In a pixelated image, more or less of the original information present in the original unpixelated version is preserved. If pixelation has been effected so that the area within which the local luminance is averaged is relatively small with regard to the whole image, it means that pixelation is carried out using a fine scale of pixelation (eg, middle picture in Fig. 3.2). In this case the relative amount of information describing the original image contents is relatively large. If the pixel size is relatively large it is a coarse-scale pixelation (eg, picture on the right in Fig. 3.2). In this case the amount of preserved information is relatively more degraded/eliminated. After pixelation, the information that has remained present from the original image is hidden or masked due to the competing interfering information added by the pixelation transform; this is in addition to the impoverishment of the original image because of elimination of the higher spatial frequencies above the low-pass cutoff value included in the original. Low spatial frequency, coarse information is less distorted and better preserved than detailed, high spatial frequency information. Consequently, any means that help to counteract the effects of the masking cues brought in by spatial quantization are helpful in restoring the correct perceptibility of the original information. Some of the first steps in this were used and demonstrated by Harmon and Julesz (1973) and Harmon (1973, 1971).

Let me list the main methods that can be used for the purpose of unmasking the contents of the original image.

1. *Filtering out the high-frequency masking cues by spatial filtering.* This method can be used in several ways. Any optical lens that helps to defocus the image at its projection plane (surface) may be used. As a result we will have a blurred image. In one case we have a natural optical lens—the one within the eye. When an object is fixated its image is projected onto the retina, with optic rays passing through the lens and converging at the retina. However, when the focal point is projected at a location nearer or farther away from the retinal plane,

Perception of Pixelated Images. DOI: http://dx.doi.org/10.1016/B978-0-12-809311-5.00004-0

the sharpness of the projected image is lost—the image is blurred. In order to blur the spatially quantized (pixelated) image one may deliberately defocus the eyes by trying to look farther than the physical image plane is or closer to this plane. A reflexive change of accommodation may help achieve this when one tries to squint one's eyes. This kind of blurring occurs with regard to the retinal plane. In a different case we have an artificial lens, such as the ones used in photocameras, TVcameras, magnifying glasses, standard eye-glasses, etc. The image projected from the original image (base image) onto a different plane (surface) can be blurred by defocusing the artificial lens. For example, when you defocus the lens of your data projector having shown a sharp image of the pixelated picture a while ago, you now see the blurred image, with initially hidden low spatial frequency information better perceived. This happens because the defocusing of the lens effectively filters out the high spatial frequencies that carry the masking noise of the edges and small corners of the square-shaped blocks of the pixelated image.

High spatial frequency can also be filtered out by other natural means pertaining to spatial resolution of the biological sensory mechanisms. Visual spatial resolution of the eye depends on the density and size of receptors and retinal ganglion cells. There is a limit in terms of how small the spatial separation between two neighboring luminous points can be so that these points can be discriminated. If the visual image is far away from the observer, the physical spatial separation between the variable luminance levels within that image cannot be projected onto clearly different receptors in the retina, which causes loss of high spatial frequency information at the retinal level. Low spatial frequency information is presented retinally though. To experience this effect, take the page depicting three images (Fig. 3.2) and move it many meters away from you. At certain distance (eg, more than 6–8 m) you perceive all three pictures as virtually the same. Fine-scale information from the mosaic of blocks of the pixelated images is filtered out and they become similar to the leftmost original picture. (In the latter the high spatial frequencies are also filtered out and the appearance of it also changes a bit, compared to when one looks at it from a close distance.) It seems also that decreasing the size of the pixelated image without surpassing the spatial resolution of the receptors nevertheless supports better identification (Bindemann et al., 2013).

The density of receptors and the concomitant spatial resolution capacity also vary between different regions of the retina. The foveal part (covering an area corresponding to about 1.5 degrees of the visual angle) has high spatial resolution capacity, but the peripheral part (receiving optical signals from more peripheral parts of the visual field) has low spatial resolution capacity. This means that even though one observes the quantized image from the same distance, either directly staring at it or attending to it "from the corner of the eye," visual appearance is very different between these two visual conditions. The peripherally presented image appears blurred, with high spatial frequency information filtered out due to suboptimal spatial resolution of the receptive system of the eye. This can be easily experienced by looking away from the image in Fig. 3.2 or by keeping the same direction of the gaze, but manually moving the picture sideways.

Artificial means to filter out high-frequency spatial information because of suboptimal spatial optical sampling capacity are manifold. A pixelated image can be photographed or videotaped from a far distance; the coarse-scale pixelation transform can be multiply applied so that a different scale value for pixelation is used for each subsequent image undergoing pixelation, so that the results of the previous pixelation are used as source images (originals, base images) in the next pixelation, but the previous resulting images are cumulatively added to (superimposed onto or blended with) the preceding images; the same coarse-scale pixelation filter is repetitively used, but applied so that variable lateral spatial shifts of the source image are introduced before each next pixelation, etc. (In the occasion of the latter-mentioned methods we have an interesting instance of a paradoxical demasking effect. By multiply applying coarse-scale pixelation transforms with varying the coarse-scale value of pixelation and cumulatively adding previous pixelation results to the subsequent ones an image is produced where the nuisance effect of the blocks of the mosaic are considerably reduced.) Moreover, the options include the often used filtering methods such as Gaussian blurring or low-pass Fourier filtering that both effectively mimic subsampling of spatial contrast.

2. *Attenuating the high-frequency masking cues by suboptimal luminous intensity and/or contrast.* The higher the SF of visual sensory cues, the more energy and the higher contrast should be used in order to pass the threshold of visibility. Therefore, if intensity and/or contrast of the pixelated image is more or less substantially decreased, HSF masking

cues lose their perceptual conspicuity or even become subliminal (ie, subthreshold of visibility), which allows the LSF content to be perceived without perturbation by HSF masking.

3. *Restricting perceptual access to the high-frequency masking cues by short exposure duration.* It is known that images presented with very short duration allow mostly coarse-scale perception and for the high spatial frequencies of the image to be well represented in the percept, longer durations are needed (Bachmann, 1991; Goffaux, Gauthier et al., 2003; Parker et al., 1997; Schyns & Oliva, 1994). Therefore, when we present a coarse-quantized image for a very short duration (eg, 1 − 30 ms) its original base image contents will be better identified compared to when the same image is observed for a longer time (eg, from 500 ms up to many seconds).

4. *Masking the masking cues.* The visual cues and features emerging in the spurious block-mosaic structure (after spatial quantization) impair the correct perception of the quantized image, despite that the coarse-scale original information is kept in the quantized image. This is essentially a masking effect. It is logical to expect that when the visual cause of the masking effect is itself masked, masking should decrease and the original image recognition improve. Indeed, Harmon and Julesz (1973) and Morrone et al. (1983) have already been successful in restoring the recognizability of the spatially quantized portraits. Harmon and Julesz (1973) attenuated spatial frequency information in the mid-range of frequencies in the quantized picture and were therefore able to restore recognizability. Morrone et al. (1983) achieved demasking by adding high spatial frequency noise to the coarse-quantized picture. Thus, a useful way to regain a more authentic perceptual impression of the original source image from the coarse-pixelated image would be to superimpose a fine-grained noise-field on the pixelated image. Random dotted noise would be just one example. A haphazard pattern of densely spaced thin lines would also be useful. The effect of demasking can also be obtained when dichoptic viewing conditions are used so that in addition to the coarse-scale pixelated image presented to one eye, a low-pass filtered version of the same image is presented to the other eye (Uttal et al., 1995b).

5. *Presenting the pixelated image in motion.* When a coarsely pixelated image is moved sideways or haphazardly around, and the speed of motion is sufficient (eg, spatial excursion with above 30 degrees/s),

recognizability of the source image is increased. Fast repetitive to-and-fro motion along the frontal plane serves this purpose well. Sensory mechanisms responsible for high-scale spatial discrimination and sensitive to fine spatial detail are more vulnerable to fast image movement than coarse-scale tuned mechanisms. Coarse-scale image elements are less distorted because of lateral smear of image elements and luminous contrast areas compared to fine-scale elements. As coarse-scale information is processed faster than fine-scale information in the brain (Schyns & Oliva, 1994; Vassilev & Stomonyakov, 1987), sampling of coarse-scale spatial modulation of contrast from a moving image is more reliable than sampling of fine-scale information. However, effective spatial frequency is increased along the motion direction due to wavelength compression (provided that pursuit eye movements are absent or are slower than image motion). Because this compression is uniform across different frequency bands, a virtual image of the original image consisting of a changed spatial frequency band is perceptually produced. In this virtual image the absolute frequencies are changed, but the relative values of the spectral components have remained invariant. This has essentially the same effect as if a coarse quantized image is observed from a greater distance. Similarly, dynamic pixelated facial images presented as video clips are recognized better than static ones (eg, Lander et al., 2001). It is possible that motion helps to emphasize the individual cues of facial identity as the invariants remaining present despite dynamic changes in the configuration of the facial cues.

6. *Slanting the plane of the pixelated image.* Take the surface containing Fig. 3.1 and slant it so that the visual angle between the image plane and the optical axis (line of sight) becomes increasingly smaller. At one point you notice that the initially difficult-to-perceive face becomes much more face-like in its perceptual appearance. While the value of effective spatial frequency increases due to the slant (more cycles from the same visual angle) and applies for all frequencies, the effect is analogous to increasing the distance of viewing the image. However, this effect works only unidimensionally and therefore can help restore recognizability when the slanting manipulation is performed along the axis including the core or the defining information for the object depicted. (An analogous caveat applies also to the method of demasking by motion.)

When is it that one may need to demask information inherent in a pixelated image? An obvious case is the forensic domain when an image that is potentially a piece of legal evidence has poor quality because of its pixelated appearance. In that case, trying to reveal more information in order to recognize a suspect would be valuable. The above-described methods may be of help. Another set of circumstances may be in relation to media politics or practice. Quite often the personal identities of individuals depicted on the printed pages of newspapers or magazines or shown in TV are masked by pixelation transform. It must be a routine check to try to see whether the person unrecognizable at first sight could be recognized after some of the above-described methods of demasking are used. The possibility of infringement of personal rights is important enough to care about and to try to avoid. A third domain relevant here is images captured and recorded from satellites or aerial photos. In this case also the image quality may be under par. If image-recognition technology cannot solve the problem of identification of a pixelated image, human observers may be helpful when using the listed "tricks" for ultimate classification, recognition, or identification.

CHAPTER 5

Explanations of the Perceptual Effects of Image Pixelation

The generic set of general explanations of why pixelation impairs perceptual identification or recognition was presented on page 43. First, higher-spatial-frequency information carrying useful cues for identification (especially the local, detailed-feature cues) is filtered out and the informational content of the original image is impoverished and degraded. Let us call this *filtering theory*. Second, the spurious new visual cues produced by the pixelation-transform mask the perceptual cues present in the original content of the image. Masking of the original image content is caused by the fine-scale edges and corners of the newly created block mosaic of the pixelated image and perhaps also by the holistic mosaic image itself. This we call *masking theory*. Third, the pixelation transform distorts the configuration of the elements of the original image. Within the coarse-scale square-shaped superpixels, the authentic configuration of the local elements and contrast gradients of the source image becomes distorted. This is the *configural distortion theory*. However, in addition to the generic and intuitive explanations more precise knowledge about the processes and mechanisms behind the perceptual degradation/masking effects of pixelation has been sought for in pertinent experimental research.

Harmon and Julesz (1973) tried to bring more clarity to the masking theory. A coarse-scale pixelated image of the face of Abraham Lincoln was used consisting of only 20 pixels per face along the vertical dimension. (This corresponds roughly to 10 cycles/image contrast modulation.) When low-pass filtering was used by eliminating spatial frequencies (SFs) above 12.6 cycles/image, recognition of Lincoln became easy and demasking effective. Then they used band-rejection filtering so that SF information closely above the carrier value for the pixelated portrait was attenuated, but high-frequency information of the pixelated image above about 40 cycles/image was not attenuated. (The band-rejection filter covered SF values between

Perception of Pixelated Images. DOI: http://dx.doi.org/10.1016/B978-0-12-809311-5.00005-2

12.3 and 39.4 cycles/image.) If masking of low-frequency facial information by the high-frequency noise of the block mosaic (ie, sharp
edges of the pixels) was the cause of the masking effect, this kind of
transform should not have eliminated masking. Actually, masking was
weak and recognition as easy as with the low-pass image. Subjectively,
the Lincoln block portrait appeared as a recognizable coarse facial
image with a grid of thin lines superimposed on it. Harmon and Julesz
(1973) concluded that masking is caused by noise that is spectrally
adjacent to the spectral contents of the recognizable coarse-scale facial
image. Masking must be mediated by the critical band of SF close to
the SF content which defines the image cues important for recognition.

At the same time, Harmon and Julesz (1973) contended that the
results of the above-described experimental manipulations cannot be
conclusive because pixelation results correlate with spectral components
of the source image, pixelation does not eliminate spatial periodicity of
the contrast change and the higher the SF, the lower the energy of the
spectral components and the possible masking effects of the high SF
(HSF) noise introduced by pixelation may be artificially attenuated.
In order to circumvent these problems, Harmon and Julesz (1973)
produced low-pass filtered images without using pixelation and masked
them by a spatially uncorrelated, aperiodic random noise with constant
average energy over the SF spectrum. The masking effect of this type of
noise depended on its SF. With noise spectral content close to the image
spectral content [ie, both having relatively low SF (LSF)], masking was
strong. When noise spectral content was much different (in this case the
band of SFs carrying noise being HSF), masking was weak. The authors
concluded again that spectral proximity of the masking noise guarantees
effective masking, thus supporting the critical band theory of masking.
Unfortunately, the effect of masking the coarse-scale, but *not pixelated
images* by spectrally adjacent random noise cannot be taken as direct
evidence in favor of the theory that the spurious high-frequency noise
produced by pixelation is the very cause of masking the coarse-scale
content of the *pixelated images*. Among other reasons, it is also doubtful
because the relative power of the HSF spurious noise in comparison with
the preserved picture-specific LSF content is very low (Morrone et al.,
1983). Moreover, the spatial orientation content of the HSF noise
produced by pixelation and the orientation content indicative of the facial
image do not overlap too much. Under these circumstances, strong
masking of LSF by HSF is doubtful.

By virtue of the methods and terminology used by Harmon and Julesz (1973), their theoretical explanation tends to be based on relatively early level sensory-perceptual processes supported by SF-tuned visual channels. A higher processing level based explanation related to perceptual organization was suggested by Morrone et al. (1983). They produced a coarse-scale pixelated version of Leonardo da Vinci's Mona Lisa consisting of 16×16 gray-level pixels. Instead of filtering out the HSF by low-pass procedures or otherwise blurring the image Morrone and colleagues added further HSF noise to the pixelated image except that the edges of the blocks were not masked by the iso-oriented noise components. Thereby they increased the spatial spread and power of the HSF predicting that if the critical-band masking theory is valid, the additional inclusion of noise should impair perception of the picture even more. However, the results showed an opposite trend: added HSF noise improved picture recognizability. Morrone et al. (1983) noted that noise destroys the block-mosaic structure, helping Mona Lisa to reemerge in perception. Theoretically, they suggest that noise effectively destroys configuration of the block portrait allowing the competing configuration of the face to become relatively more dominant. The spatial stratification of local oriented elements in the image can suggest one or another more global organization of the image and when the local cues suggesting the mosaic structure are attenuated, the less attenuated local cues suggesting a facial structure acquire the upper hand. Yet, because using noise that literally masks the local HSF cues of the blocks' edges also restores the original image visibility, the masking theory cannot be completely disregarded.

It is necessary to give a comment on the use of the concept of "masking" in the present theoretical context where the impairment of perception of object images with varying SF content is dealt with. This term can be used for interactions that may take place at different levels of the sensory-perceptual system. For example, "masking" may refer to inhibitory interactions between channels tuned to certain SF content of the stimuli independent of where in the image space this SF content originates from. On the other hand, "masking" may refer to inhibitory interactions or decrease of signal-to-noise ratio caused by relative location of the masking and masked image cues in the 2D image space. Thus, it is possible that SF spectra of the masking and masked image are not overlapping or even very close, but the respective visual cues overlap or are immediately adjacent in space. It is also possible that the masking and masked cues are not precisely spatially nested so as to overlap within the 2D image space, but they are overlapping precisely in terms of their SF spectra. In other words, the concept of masking itself is too ambiguous for advancing our theoretical understanding unless we specify what we mean by this term. An important criterion for specifying masking is spatial correlation between the masking and masked image cues carried by their SF content.

Additional data useful for the theoretical understanding of face-masking were presented by Caelli and Yuzyk (1985). They prepared versions of two test faces by filtering them with two-dimensional annulus-shaped Gaussian filters systematically changing the SF value of the passed SF. (The filtering value was varied from 0 to 64 picture cycles in five steps, producing five versions of filtered images for each of the two original pictures of faces.) A 5×5 matrix of face–face montages was prepared so that each of the face 1 filtered versions was combined with each of the face 2 filtered versions. Caelli and Yuzyk (1985) were interested in what the outcome would be of such face-by-face masking in terms of the effects of SF content on three possible perceptual outcomes: (1) perception of two, separate and distinguishable faces, (2) perception of one, new, fused image, (3) perceptual dominance of one of the faces which masks the other. The basic results were instructive with regard to our problem of ambiguity of the concept of masking. Perceptual segregation of two faces occurred when the spectral distance between the two filtered images in the montage was larger than 3 octaves of SF (peak-to-peak measurement). Masking was strongest with intermediate spectral distance between the two images in the montage (2 or 3 octaves). Fusion occurred at a comparable level with respective response percents decreasing from about 50% to about 30%. In the control conditions filtered face images were combined with filtered texture and scene images to observe the effects of different types of SF-filtered content spatially not highly correlated with face images. In general, perceptual segregation increased with spectral separation of images, being higher in absolute terms for face–texture and face–scene montages than face–face montages. Masking of faces with textured noise decreased with an increase in spectral distance between the images. Masking of faces by scenes was maximal with intermediate distance between image spectra, similarly to what happened with face–face masking. Fusion of two images into one perceptual entity did not depend strongly on spectral similarity. When perception of montages with identical SF-filtering value were combined, fusion was considerably dependent on spatial correlation between the components of the images. Caelli and Yuzyk (1985) conclude that masking is not necessarily determined by the spectral differences between the images, but is more strongly influenced by their spatial correlations. Instead of, or in addition to masking, interaction of two images may lead either to segregation or fusion, the latter being more likely with high spatial correlation between the images. Importantly, large enough SF spectral differences also mean that image cues are more clearly separable also in terms of their *spatial decorrelation.* Thus, similarity between target image and mask image spatial structures seems to be the main determinant of the masking effects.

The importance of the relative spatial nesting of the visual cues carried by different SF contents of the object image is demonstrated also by Morrone and Burr (1997). We remember that by the blocking transform, spurious HSF cues of the block-mosaic are introduced to the image in addition to the original spectral content. Morrone and Burr (1997) applied a phase-shifting operation selectively to the spurious HSF cues of the block-mosaic image. (The spurious HSF cues were extracted elegantly by subtracting the low-pass filtered image from the block-mosaic image. Phase-shifting was then applied to this residual image. Phase-shifting did not change the location or magnitude of the blocks' edges related HSF information because the magnitude of the contrast is the same, but with opposite sign when $\pi/2$ shift is used.) The resulting new image was much easier to recognize than the initial pixelated image. Subjectively, the coarse LSF image appeared in transparency behind a grid structure of the lines. Critical-band masking theory could not explain this result. The results were explained by the local energy model of image organization developed by the authors. The features of the image are calculated separately at each scale in the initial step for the buildup of the image description in the brain. This computation is based on the peaks of local energy. Separate feature maps are recombined proportionally with scale size and summed. High-scale operators are automatically favored and with spatially quantized block-mosaic images, the high-frequency block structure dominates the *combined* map. LSF information becomes unavailable as an *independent source* of perceptual information. Image features are by definition located at points of phase congruence of the harmonic components. When this phase congruence is taken away by the artificial means, as it was used by Morrone and Burr (1997) when HSF spurious noise was phase-shifted, block-mosaic dominance was attenuated and the coarse-scale cues of the original image were more easily perceived. [For the sake of completeness consider also the comment by Lu and Sperling (1996), who—in a difficult-to-follow argumentation—posit that the Lincoln picture perception problem does not exist.]

The results of Morrone et al. (1983), Morrone and Burr (1997), and Caelli and Yuzyk (1985) suggest that spatial decorrelation of the structures of the original image and the spurious cues of the pixelated mosaic image set the stage for perceptual competition between two wholistic interpretations of the image structure. If some means can be used for biasing this competition in favor of the coarse-scale original image content extraction—for example, by adding noise to the pixelated image or filtering out HSF components from it—then the pixelated original image can be demasked.

The importance of the spatial stratification of the elements of an image independent of its SF content is also stressed by another observation. If a line drawing of a face based on HSF information is spatially quantized at a coarse-scale level (ie, pixelated coarsely), the resulting representation consisting of spatially organized relatively large local blocks nevertheless allows easy recognition of the face (Sergent, 1986). This means that the visual system can produce LSF channel-based representations from the HSF-carried input information. However, as noted by Sergent (1986), this can happen when the HSF cues defining the image are not spatially densely arranged. Their spatial separation should be larger than the basic cycle of the SF filter applied to the image.

Explanations of the masking effects brought about by pixelation of the source images have often assumed the existence of independent SF channels tuned to different frequency bands. Suppressive interactions between these channels are taken as the cause of masking. However, the results of Morrone et al. (1983), Morrone and Burr (1997), and Caelli and Yuzyk (1985) put this interpretation in doubt. As was nicely put by Morgan and Watt (1997), the sampling failure involved in perceiving a pixelated image can be explained if we accept that by applying the spatial blocking transformations, peaks in local energy in LSF channels are distorted. This assumption requires nonindependence of the access to LSF channels in building up the perceptual object experience.

In the MIRAGE model of spatiotemporal image processing also the early visual channels are of a "non-Fourier" nature. According to this approach, positional noise is the limiting factor for spatial discrimination, including the perception of spatially quantized images.

Between-channels masking is questioned also by some additional novel experimental findings. Uttal et al. (1997) refer to the effect found by Durgin and Profitt (1993). When these researchers added thin lines to the pixelated image so that all the edges of the blocks were emphasized by increasing their contrast, masking was reduced instead of the expected increase predicted by the between-channels theory. Subsequently, Smilek, Rempel, and Enns (2006) used a similar method, overlaying a mesh screen on a pixelated gray-level image subtending 12 degrees. Compared to the blurred appearance of the pixelated image, the same image when covered by mesh (with its wires either covering the pixel edges or being shifted), appeared to have higher clarity. When the original unpixelized images were covered by the mesh screen, their apparent clarity decreased. For example, when a picture included a human face, the face part subtended a region that was pixelated into about 10−30 pixels unidimensionally. The illusion of clarity was definitely present in the conditions of relatively fine-grained pixelation, but not with the coarsest levels of pixelation. This illusion was interpreted as an image segmentation process that falsely attributes the edges of the quantized blocks to the screen. Another paradoxical result was reported by William Uttal himself (Uttal et al., 1995a). In the original Harmon and Julesz (1973) study showing demasking by filtering, the pixelation (blocking) of the Lincoln picture was applied first, followed by SF filtering. According to the critical-band masking theory, when these degradations would be applied in a reversed order, masking should increase instead of decrease. This is expected because blocking transform adds higher SF noise to the image. Actually, Uttal et al. (1995a) demonstrated a decrease in masking. This again speaks against the critical-band masking theory.

Uttal and colleagues list several arguments for why the critical-band masking theory cannot be correct or at least sufficient (Uttal et al. 1997). (1) For small images of faces, either order of applying the types of degradation—blocking and SF filtering—produces enhanced recognition. (2) The power of high frequencies is insufficient to mask the LSF content which has much higher power. (3) Adding power to HSF content at the edges of the blocks in the quantized image does not impair recognition. (4) Reversal of the phase of the edges of the blocks helps overcome the masking effect of spatial quantization (Morrone & Burr, 1994). (5) The effects of degradation depend on the subjects' task (see the following parts of the book). Therefore, low-level interchannel interactions are not sufficient to explain the masking effect obtained by pixelation. (6) Adding random visual noise to interfere with HSF processing may enhance recognition (eg, Morrone et al., 1983), but may also decrease it (Uttal et al. 1995a, 1995c). (7) There are several empirical facts showing that perceptual organization of the image cues in 2D or virtual 3D take priority over the SF energies in determining the success of recognition. Generalizing on the work of his own research team and the facts gathered from research literature, Uttal asserts that models based on single stimulus attributes cannot explain the masking and demasking effects in perceiving pixelated images. Instead, a more eclectic set of different mechanisms involved may be a better theoretical perspective (Uttal, 1998; Uttal et al., 1995a, 1995c, 1997).

More data speaking against the critical-band masking and its dependence on the energy of the spurious spectral components have also been gathered in subsequent research (eg, Sperling & Hsu, 2004; Bachmann, Luiga, & Põder, 2004, 2005a, 2005b). These data will be discussed in relation to the problems of time course functions of perception of target images and alternative explanations of the effect of pixelation in the later parts of this book.

To shed more light on the straightforward, simple question about whether pixelation effects on recognition are equivalent to spatial-frequency Fourier filtering, Kristjan Madisson (2010) carried out a trivial experiment in our lab. Recognition of familiar faces was compared between two conditions: pixelation transform with 18, 12, and 8 pix/face levels of pixel resolution and SF filtering with 9, 6, and 4 cycles/degree smooth filter. Results were clearcut—pixelation impaired recognition more than SF filtering ($P < 0.0001$). The average proportions of correct recognition were 0.31 and 0.50, respectively. This applies to all levels of image impoverishment. The effects of pixelation cannot be limited to simple SF filtering.

CHAPTER 6

Limits and Optima of the Parameters of Pixelated Images for Perception

6.1 HOW MANY PIXELS CAN CARRY USEFUL INFORMATION FOR CORRECT DISCRIMINATION?

In the seminal work by Harmon and Julesz (Harmon, 1973; Harmon & Julesz, 1973) only one value of coarseness of pixelation was used. The picture with the Abraham Lincoln portrait was blocked to contain 14×19 blocks (horizontal and vertical dimensions, respectively). The area in the picture depicting the head and face of Lincoln was even smaller—about 11×14 pixels. With this coarseness level of pixelation naïve observers have difficulty in recognizing the picture. Unfortunately, Harmon and Julesz did not change the level of pixelation parametrically so as to measure the thresholds or rate of correct recognition depending on the scale of pixelation. In the following study by Morrone et al. (1983) a 16×16 pixelated area including the Mona Lisa portrait was used, with facial area within the picture measuring about 10×14 pixels; no attempt was taken to systematically change the coarseness level of pixelation to see the resulting effects on recognizability. The early published study, where coarseness of pixelation was varied to see the effects of this manipulation, used only three levels of pixelation—128×128, 56×56, and 36×36 pixels applied to an image of the human eye and an image of a human face (Bachmann, 1987). Recognition dropped significantly with an increase in coarseness. (Exposure duration was varied between 1, 20, and 1000 ms. With 1000 ms duration 128×128 pixels images were recognized perfectly, 56×56 pixels images at about 65% correct recognition, and 36×36 pixels images only at about 20% correct recognition level.) However, because in the object recognition task pixelated images were included among 10 alternatives where the majority of the images depicted unpixelated images of objects and as the 32 observers did not know beforehand anything about the fact that pixelated pictures might be used, the results are not very informative about the recognition thresholds for pixelated object images. Also, using

Perception of Pixelated Images. DOI: http://dx.doi.org/10.1016/B978-0-12-809311-5.00006-4

only three levels of pixelation is not very informative about the effect of pixelation coarseness on recognition.

In the following study Bachmann (1991) varied pixelation level more systematically and used a task where subjects knew they had to identify human faces from their pixelated images. Six alternative gray-level source images depicting frontoparallel views of male faces were used, each pixelated at eight levels of pixelation: 15, 18, 21, 24, 27, 32, 44, and 74 pixels per face width (pix/f, horizontally). Thus, the stimulus set consisted of 48 pixelated images and observers were asked to identify each presented pixelated picture by referring to the original unpixelated photograph corresponding to the pixelated version. Exposure duration was varied between six values—1, 4, 8, 20, 40, and 100 ms. Correct identification increased abruptly with increase in exposure duration from 1 to 4 ms or 8 ms (depending on pixelation level) and then this increase slowed. Most importantly, images pixelated at different levels of pixelation coarseness were identified with a comparable rate of correct responses, except that the most coarse pixelation at 15 pix/f abruptly decreased identification accuracy. Thus, there seems to be a discontinuity in the effect of level of pixelation so that faces pixelated at 18 pix/f or more are reasonably identified, but changing pixelation level down to 15 pix/f dramatically impairs forced-choice identification. (Remember that 15 pix/f roughly corresponds to 7.5 cycles of contrast per face and 18 pix/f to 9 cycles per face.) In order to prevent undesirable side-effects and artifacts of image blurring by defocusing eyes or by averting gaze, Landolt visual acuity stimuli were randomly inserted into the succession of facial images, with subjects instructed to discriminate these Landolts.

Why does face identifiability increase considerably when instead of the 15 pix/f images, 18 pix/f images are used? Increasing local feature discrimination cannot be the cause of this effect for a simple reason: 18 pix/f is insufficient for depicting any recognizable facial feature as such. For example, an eye in this kind of block portrait consists only of about 3×6 square-shaped pixels, a mouth about 3×7 pixels, and the nostrils part of the nose about 2×5 pixels. If any of these local pixelated areas were to be cut out and shown separately to observers it would be impossible to recognize any facial element or feature. The remaining main explanations are (1) allowing the critical band of facial spatial frequency (SF) content of the coarse-quantized faces to be represented after reaching 18 pix/f level of quantization; (2) allowing a wholistic configuration of facial element tokens to be veridically represented after reaching 18 pix/f level of quantization. However, the first of the above explanations seems not to be the best option if we remember the results of the studies by Caelli and Yuzyk (1985), Morrone and Burr (1997), and Morgan and Watt (1997), all stressing the crucial role of spatial correlation of image cues pertaining to different spectral components and a relatively less important impact of the spectral overlap as such. Perhaps then the step from 15 to 18 pix/f pixelation value produces favorable conditions for local energy spatial distribution analysis so that the output of this processing stage makes contact with a certain face configural template without being sufficiently perturbed by the results of the analysis of spatial distribution of the block mosaic local energies. To put it in a perhaps simplified, but not unreasonable, way: with 15 pix/f spatial quantization the perceptual interpretation of the ambiguous visual object structure (mosaic of squares vs face) is biased towards a mosaic while with 18 pix/f (and above) the structure of blocks biases interpretation towards a face.

Costen et al. (1994, 1996) investigated face recognition with three pixelation levels in one experiment (six faces were spatially quantized at 11, 21, and 42 pix/f images each and presented for 100 ms) and with four pixelation levels (9, 12, 23, and 45 pix/f) in additional experiments. An abrupt drop in performance was observed when there were less than 21–23 pixels per face, that is, with 9, 11, or 12 pix/f images perception of facial identity was difficult. All these pixelation values leading to low accuracy of recognition or discrimination were lower than the critical pixelation value in Bachmann (1991), thus supporting the earlier results. The range of variable pixelation coarseness values was extended even more by Bachmann and Kahusk (1997). They used 9, 10, 11, 12, 13, 14, 16, and 96 pix/f images of eight frontoparallel male faces. Pixelation at 16 pix/f and less resulted in a low level of identification at about 50% correct and less compared to the 75% of the 96 pix/f condition. Again, the equivalent of less than 8–11 cycles/face spatial contrast (ie, less than about 20 pix/f) appears suboptimal for accurate perceptual identification.

In the work discussed so far there seems to be a consensus about the critical value of pixelation. For facial images the critical coarseness scale of pixelation beyond which further decrease in the number of pixels/face causes an abrupt impairment of identification equals about 18−21 pix/f. Notably, this is roughly the equivalent of the SF filtering value which was found to be optimal for face identification by Gold, Bennett, and Sekuler (1999). These researchers presented 10 alternative band-pass filtered images for about half a second, embedded in 2D dynamic Gaussian noise. Good identification of faces was possible with filter center frequency at 8.8 cycles/face. This is roughly equivalent to the 17.6 pix/f pixelation value. (Comparing the results of human observers to the ideal observer data, Gold et al. concluded that the failure to identify faces from suboptimal spectral content of the filtered images reflects constraints on visual processing rather than the absence of information specifying a stimulus.)

In most of the studies of the critical range of pixelation-scales for recognition or identification, a relatively small number of alternative images has been used. Typically, from four to eight alternatives are prepared for identification or the recognition task. In "real-life" settings of human perception, as well as in the attempts to develop machine vision systems, which are both more important to study in the applied research context, more often than not the potential pools of alternatives are quite larger. Therefore, the experimental results reported by Bhatia, Lakshminarayanan, Samal, and Welland (1995) where they used 92 alternative images of faces as the original images for degradation are especially valuable. Bhatia and colleagues used a two-alternative forced-choice paradigm where pairs of images were shown for 500 ms each and observers had to decide which of the two images in a pair depicts a face. All individual images subtended 3.56 degrees of visual angle from the observer's point of view. Human faces (termed target images) were paired with nontarget images, such as mammalian nonhuman faces, scrambled human faces, and inanimate objects. Two experiments were carried out. In the first experiment each observer in the sample of 43 subjects was given about 5000−6000 pairs for decision. Pixelation coarseness level was varied between five levels: 8×8, 16×16, 32×32, 64×64, and 128×128 pixels defining the image (pix/im). Bhatia and colleagues also varied the gray-level scales over which the blocks of the pixelated images were varied (2, 4, 8, and 16 gray-level scales were used). While the accuracy level of face discrimination for pixelated images was generally very high (between 90% and 100% correct discriminations), the abrupt decrease in accuracy was observed with the most coarse pixelation level at 8×8 pix/im for all gray-level conditions. (The condition with only two gray levels produced the worst results, with only less than 20% above chance responses.) The very high accuracy of discrimination with facial images consisting of 16 pix/f is somewhat different from what has been found by Bachmann (1991) and Costen et al. (1994, 1996). In these studies changing spatial resolution of the pixelated faces down from 18 to 20 pix/f coarseness level values already produced an abrupt decrease in identification performance. The main reason for this discrepancy is most probably related to the different tasks used by Bhatia and colleagues on the one hand and other researchers on the other hand. In Bhatia et al. (1995) a face had to be discriminated from a nonface image in a pair of simultaneously presented pictures, whereas in other studies more subtle discriminations had to be made between different faces.

In the second experiment, Bhatia et al. (1995) extended the pool of pixelated images in order to more precisely analyze the decrease in visual discrimination with a systematic change in pixelation coarseness level. Altogether, 18 scale levels of pixelation were used, beginning with the coarsest at 4×4 pix/im and ending with 32×32 pix/im spatial quantization; eight gray levels were used, which in combination with pixelation levels determined that each original picture generated 144 pixelated images. With 36 observers a total of more than 205,000 trials were administered. Because correct responding by chance would be 50% in this experimental design, an operational value for threshold of perceptual discrimination could be set at 75%. A probit analysis showed that the more gray-scale levels were used, the smaller the number of pixels per image necessary to reach the 75% correct response level. The 75% correct response threshold dropped from about 11 pix/f when only one gradation of gray level was used to about 6.5 pix/f with three gray levels. If more than three or four gray levels were used for image depiction, no further decrease in the number of pix/im defining the 75% correct level of performance was observed. Thus, when gray-level facial images are pixelated, about four gray levels for defining the blocks with different intensity would be sufficient, provided that the range of contrast was large enough. Bhatia and colleagues plotted the percentage of correct responses as a function pixelation level, with gray level as a parameter. The fastest increase in correct discrimination from near-chance performance to about 80–90% was observed when the pixelation level value reached 7–8 pix/f. Bhatia and colleagues suggest that by taking the 75% criterion as the guiding principle, 7×7 and 8×8 pix/im could be used as benchmarks for evaluating the performance of a face discrimination system, whether natural or artificial. (Here it is important to remember that these specific values refer to a specific task of discrimination of human faces as such from simultaneously presented nonfacial, nonhuman facial, or scrambled facial images. For other tasks the benchmark values and values of image degradation leading to an abrupt loss of veridicality of perception may be different.)

Usually the original pictures of faces from which pixelated versions were produced depicted internal facial elements as well as the overall shape of the face, often including the hairline. It is therefore difficult to know whether the overall shape-related information or information about the internal facial features and their configuration is what suffers from pixelation the most. In this respect the experiments by Uttal et al. (1997) were valuable because they used frontoparallel gray-level facial images (3.5 degrees horizontally) by eliminating the extraneous cues except the facial features. For this, the stimulus faces were cropped by framing the face image with a standard cutout template which guaranteed that details of hairstyle and hairline as well as the head shape were removed. Four levels of pixelation were used producing spatially quantized images with 10.5, 14, 21, and 42 pix/f spatial resolution (approximate values). Exposure duration was 100 ms. Twelve-alternative forced-choice identification tasks were used. While the two finest levels of pixelation (21 and 42 pix/f) produced almost perfect identification, the 14 pix/f resolution caused the percentage of correct responses to drop by more than 10% and the 10.5 pix/f added a further similar decrease. The critical value equal to 14 pix/f corresponds well with similar earlier data (Bachmann, 1991; Costen et al., 1994). However, despite that the task in the Uttal and co-workers' investigation had to be more difficult than those used in the preceding studies, the overall performance was at a surprisingly high level. This is difficult to explain when we remember that in those other studies the accuracy levels of performance with images pixelated close to the 14 pix/f varied between 40% and 60%. Probably the differences in stimuli in terms of their mutual similarity, contrast, and how well learned they were must have caused the differences in the absolute level of accuracy.

It should be expected that the critical value of pixelation leading to a drop in performance is related not only to the number of alternatives, familiarity of the stimulus objects, task difficulty, and other procedural factors, but depends also on the types of stimulus objects. Although faces have been by far the most popular types of images, other objects have been also used. Uttal et al. (1995a, 1995c) pixelated 12 small solid silhouettes of aircraft subtending approximately 1 degree vertically. In one of the experiments six levels of blocking were used rendering images with approximately 22, 15, 11, 9, 7.5, and 6.4 pix/im resolution. (Notice that only original solid images are not gray scale. When a small light shape is placed on a dark background, coarse-scale pixelation introduces nonhomogeneities in the luminance level of pixels because more or less of the dark background luminance value is sampled for defining the averaged luminance of the pixel. This effect is larger, the closer a pixel is to the edge of the solid shape.) Observers were instructed to specify whether the two sequentially presented 100-ms stimuli were the same or different. The chance level of correct performance is therefore 50%. Independent of the level of pixelation, the matching performance was almost perfect at more than 90% correct responses. The two main reasons why this task was so easy and very coarse pixelation levels at about 7 pix/im and less were not effective at degrading are: (1) the shape primitives of the silhouettes remain largely the same despite various levels of pixelation; (2) the task of comparing two different successive images separated by about 1 s in time for assessing their identity is perhaps a too-easy task. Moreover, performing this task may well be accomplished by attending to visual cues that are not diagnostic for correctly perceiving the category or identity of the object, but are simply some spurious visual cues for discriminating the shapes, whatever their content.

Uttal and colleagues were aware of the limitations of the discrimination task and in a subsequent study they used an identification task (Uttal et al., 1995a). In this identification study, Uttal and co-workers were primarily interested in what effects different combined degradations may have. Nevertheless, they included also a control condition for measuring the baseline effects of blocking (pixelation) degradation. This is what we present here. The same 12 aircraft images as in the preceding study were used, pixelated at 13, 7.8, and 5.4 pix/im levels of pixelation along the horizontal dimension. Observers were presented with 100 ms single presentations of the images and asked to identify them by choosing one of the 12 response keys corresponding to the alternative images. Images consisting of 13 pixels horizontally were identified virtually perfectly; 7.8 pix/im resolution yielded less than about 85% correct responses whereas 5.4 pix/im resolution produced slightly more than 66% correct responses. A similar dynamics of the decrease of performance in a discrimination task from about 80% correct to about 65% correct (with 50% correct guessing level) was found with the same stimuli in the block control condition also in another paper (Uttal et al., 1995b). We see that compared to faces, images of the silhouettes of artifacts tolerate a bit more coarse spatial quantization for veridical perceptual processing. While with gray-scale faces most of the researchers have found the critical range of pixelation at about 15–20 pix/im horizontally, in the experiments by Uttal's team, aircraft images pixelated at finer levels beyond about 8 pix/im are already well identifiable or discriminable. The best candidate explanation for this discrepancy owes to two principal differences between the artifact silhouette images and faces. The former are well discriminable by their outline shapes, while the general outline shapes of faces are quite similar. Also, the configurations of the elements of individual faces more or less correspond to the generalized face prototype (face template) with relatively small inter-individual differences, which necessitates not too coarsely pixelated images in order to allow discrimination. Configural cues of the artifacts used by Uttal and colleagues varied so widely that this may have allowed discrimination at a more coarse level of pixelation. [Somewhat similarly, the critical band of spatial frequencies for letter-shape identification compared to face identification was observed by Gold, Bennett, and Sekuler (1999). Although letters could be identified at center frequencies of filtering below 6.2 cycles/letter, this low level of filtering caused faces to become effectively unidentifiable. If we look at silhouettes of artifacts then certain similarity to letter shapes may be noticed. Unfortunately, Gold and associates did not use pixelation transforms.]

In the study by Lander et al. (2001) pixelation value was varied only between 10 and 20 pix/f. Videoclips with moving and static images of 20 faces were shown for 2.5 s and subjects were instructed to recognize the depicted faces. Mean hit rate for 10 pix/f static faces decreased from 52.5% when the size of the face was 3.5 × 3.7 degrees to 45.8% when the size of the facial image was decreased to 1.8 × 1.9 degrees. The effect of changing the size of the facial image was reversed with 20 pix/f static faces: hit rate was 66.7% with 3.5 × 3.7 degrees and 75.8% with 1.8 × 1.9 degrees images. The optimum number of pixels per face depends on the size of the facial images. Compared to static images, dynamic clips depicting pixelated images allowed face recognition to increase by about 10%.

Pixelated images presented for identification or recognition may be captured from full-face original source images including head shape, hair, and facial hair if present or they may depict only the internal part of faces with a configuration of basic facial elements such as eyes, nose, and mouth. Sinha et al. (2006) paid attention to comparing the effects of image degradation with full-face and internal-face images. In addition, they also lowered the random guessing level and the possible effects of nonfacial confounding cues of images by using images of celebrities without subjects knowing beforehand who would be shown. Color images of 36 celebrity persons were shown at eight levels of pixelation: 150×210, 37×52, 19×27, 12×17, 9×13, 7×10, 6×8, and 5×7 pixels per face. Even with the finest pixelation at 150×210 pix/f recognition was barely above 80% correct responses when configurations of internal parts of faces were shown. An abrupt drop in correct responses was found from 70% to slightly above 30% when pixelation changed from 37×52 to 19×27 pix/f. Internal parts configuration allowed recognition just at close to a chance level with 12×17 pix/f. The critical zone of decrease of performance generally corresponds to the data gathered by Bachmann and Costen: 19 horizontal pixels per face and/or less leads to face-recognition deterioration. We should notice that in the above-mentioned earlier pertinent studies, extraneous cues for recognition were controlled and unfamiliar faces were used. This puts special importance on the configural facial cues for recognition and identification, which in turn makes the earlier studies and the experiments by Sinha et al. (2006) mutually compatible. When well-known full-face images of celebrities with their characteristic head shape, hairstyle, and configural facial features were used, the general level of correct responding significantly increased and any critical values indicative of an abrupt drop of performance were not apparent. (A very slight decrease in correct responding from a perfect level was found when pixelation reached 19×27 pix/f. At the pixelation level equal to 12×17 pix/f where usually a drop has been found, faces were still recognized at about the 90% correct level. Even with 5×7 pix/f correct responding reached almost 30%.) Sinha et al. (2006) stress that the overall head configuration, together with hairstyle may be an important cue for veridical person recognition from a degraded image.

Perhaps interactions between the external shape and style cues with internal configural cues of the facial elements are the important mode of the work of face-recognition algorithms implemented by the human visual brain systems. Importantly, these researchers also used a control condition with internal local features depicted without being arranged in the correct configuration. This condition allowed fairly good recognition with 150×210 pix/f, assumingly allowing local characteristic features to be communicated by a fine-scale image pixelation. With 37×52 pix/f recognition dropped to a mere 30% and a further decrease in the number of pixels brought recognition at a chance level close to zero. Altogether, the results obtained by Sinha et al. (2006) emphasize the importance of the wholistic and configural processes in efficient face recognition and demonstrate high tolerance of these gross facial cues to substantial image degradation. At the same time the facial image configurations of unfamiliar persons may be much more vulnerable in terms of the effects of degradation brought about by the pixelation transform.

The importance of appearance cues other than just the internal parts of a face was stressed also by Demanet et al. (2007). They presented 90 pictures of famous people and 90 pictures of unknown persons pixelated at three levels of coarseness: 8, 14, and 20 pixles per face width. However, pictures of these people depicted a wider visual context than just a face; whole busts and whole head area including hairstyle and head shape were captured. Pixelation masking area covered either only the face, the whole head, or the whole bust. Images were presented for 3 s and each person was shown to each observer only once, which provided a useful control for the repetition-based effects of expectancy, reliance on spurious nonidentity-related cues, etc. The results showed the following. The overall level of recognition of pixelated images where only the face was masked was $d' > 1.0$, corresponding to more than 45% correct recognition. This showed that face pixelation cannot be taken as a guaranteed method of masking nullifying all capacity for person identification. In the 8-pix/f condition, where only the face part of the images was masked by pixelation, d' equaled 1.0. The same level of pixelation with the whole head masked produced d' approximately 0.5 and the whole bust masking led to d' of about 0.25. Clearly, person identification is substantially supported not only by facial cues, but also by the head and bust level information. In the 14-pix/f condition the respective d' values were approximately: 1.6, 1.5, and 1.25. In the 20-pix/f condition the d' values were, respectively, 2.2, 2.3, and 2.3. This pattern of results shows that additional context besides face as such (which is solely masked) becomes especially important when masking by pixelation is executed at the coarse level, but not so much when pixelation allows some finer-scale information to be kept intact.

Whole-face images were also used by Bindemann et al. (2013). They used a face-matching task where original faces sampled from the pool of 160 facial images of unfamiliar persons were paired with their pixelated versions. Subsampling for pixelation was performed with Adobe Photoshop Mosaic function, which transforms images into isoluminant blocks by weighted averaging of the luminances of all pixels in the block. Subjects had to produce same–different matching responses for these pairs where half of the trials depicted the same persons (ie, 50% correct responses can be produced by chance). Pixelation level varied between 8, 14, and 20 pix/f horizontally. Stimuli were visible until a response was made. When the faces in pairs belonged to same persons, correct matching responses dropped from 90% with original images paired to about 66% with 20 pix/f, less than 60% with 14 pix/f, and even less than 50% with 8 pix/f. Similarly to earlier identification research, 8 pix/f appears too much degraded for more or less successful perceptual processing. Surprisingly, the proportion of correct responses for mismatch pairs was at an equal level—about 65%—with all three levels of pixelation. When the image size of the 20 pix/f faces and their original companions in pairs was reduced to one-half, one-fourth, and one-eighth of the original size, the correct matching response rate increased. Decreasing the image helps filter out high-SF content of the masking cues of the blocks and prevail of the configural wholistic information.

Effects of pixelation on reading the written text (Arial font Helvetica) were studied by Pérez Fornos, Sommerhalder, Rappaz, Safran, and Pelizzone (2005). A 100-word text was displayed word by word in a 10×7 degrees window. This window size allowed five letters to be simultaneously visible in the window. The number of words correctly read was used as a dependent measure. Words were depicted at five levels of pixelation: 28,000, 1750, 572, 280, and 166 pixels per window. The rate of presentation was controlled by feedback from eye movements. Performance began to drop from above 95% correct to about 76% correct when instead of 572 pixels (c. 154 pix/letter), 280 pixels

(c. 56 pix/letter) carried letter and word information. A further drop to about 30% correct followed when 166 pixels (c. 35 pix/letter) resolution was used. Thus, less than 10 pixels per letter unidimensionally seems to cause a definite drop in reading performance. This estimate is again close to the critical pixelation or spatial filtering values of nonfacial images reported by Uttal and colleagues and Gold and colleagues.

Thus far we have reviewed the studies where the optimal pixelation values have been explored with facial, aircraft silhouette, and letter stimuli. However, both the theoretical and applied interests are concerned also with other types of images. This is especially obvious with images of scenes and landscapes and images used for visualization where large amounts of numerical data are communicated to experts or end-users graphically. It is of course interesting for theoretical purposes to use pixelation of stimulus images in order to reveal the information-processing algorithms and regularities involved in scene perception by human subjects. It is not less interesting for practical purposes to know the limits of veridical perception set by the level of pixelation. On the one hand, we want to know the upper limits of degradation still allowing a satisfactory performance by operators in the man—machine systems or reliable performance by eyewitnesses based on pixelated images. On the other hand, if the known limits set by pixelation permit using more coarsely pixelated images without a cost in performance, the technical and financial economy in developing and building communication systems would be substantial. In some cases using pixelation, while affecting performance depending on coarseness, does not harm the extraction of the gist of an image too much when certain critical value is surpassed (Torralba, 2009). This regularity is related to scene categorization. In some other cases block resolution does not have a significant impact on task performance (Borgo et al., 2010). This applies to data visualization. Another important aspect in applied vision concerns the dynamics of eye movements. Are there any data on how pixelation affects image processing as mediated by visual fixations?

Natural images were pixelated at eight levels by Judd, Durand, and Torralba (2011). Beginning with images depicted by four pixels vertically, the pixelation level was increased by doubling each previous level value so that the finest pixelated images had 512 pix/im vertically. Thereafter, these images were upsampled to the original pixelation level at 860 × 1024 pix/im. Judd et al. (2011) compared eye fixations between different observers and between different pixelation levels as a function of image degradation. They found that fixations recorded with lower-resolution images can predict fixations on higher-resolution images. The typical bias to fixate central parts of images is augmented with lower-resolution images. The consistency between fixations increases up to the level between 16 and 64 pixels/image depending on images and other conditions. The simpler the image, the more consistent are the fixations.

6.2 HOW PERCEPTION OF PIXELATED IMAGES DEPENDS ON THEIR DURATION

In addition to spatial variables, temporal variables also affect pixelated image perception. In most of the studies, invariant durations or variable long durations of pixelated images are used. Typical values of how long a target picture is displayed are, for example, 100 ms, 500 ms, or many hundreds of ms until a subject responds and the image is switched off. Only a few studies have used different target-stimulus durations, which allows for the exploration of temporal limitations in and temporal formation of the perception of pixelated images.

Three different durations and three different pixelation levels were used by Bachmann (1987). Square-shaped gray-level images subtending approximately 6−10 degrees of visual angle (from the observers' point of view) were spatially quantized versions of a human eye and a human face. These pixelated images consisted of 36×36, 56×56, or 128×128 pixels. They were presented for recognition among eight unpixelated images otherwise irrelevant for the study except that when the majority of stimuli-alternatives were typical gray-scale pictures of objects (1) observers would not compromise visual acuity in order to perform well in the recognition task and (2) they would not know that specifically pixelated images were the main interest of the experimenters. There were 288 observers, 32 in each of the 9 groups specified according to the pixelation coarseness value combined with stimulus duration value used for that group—36^2 pixels and 1 ms duration, 36^2 pixels and 20 ms duration, 36^2 pixels and 1000 ms duration, 56^2 pixels and 1 ms duration, 56^2 pixels and 20 ms duration, 56^2 pixels and 1000 ms duration, 128^2 pixels and 1 ms duration, 128^2 pixels and 20 ms duration, and 128^2 pixels and 1000 ms duration. Thus each group received 10 presentations, 8 unpixelated objects and 2 pixelated objects. This procedure also helps avoid any learning effects because each stimulus-image was seen only once by a subject. Subjects had to respond by writing down the category of an object they thought they saw. A forced response procedure was used requiring subjects to guess when they did not recognize the object from a brief exposure. The likelihood of correct guessing was very small, remaining below 0.05.

The results of this experiment were rather intriguing. Let us begin with more or less expected results. When pixelation level was fine-scale (128^2 pix/im) allowing image detail to be represented relatively well, correct recognition increased monotonically with exposure duration (about 60%, 90%, 100% correct with 1, 20, and 1000 ms, respectively). When the intermediate coarseness level of pixelation was used (56^2 pix/im), 1 ms exposure duration led to 40% correct recognition, 20 ms brought a further increase in correct recognition (above 60% correct), but a considerable further increase in exposure duration (up to 1000 ms) did not increase recognition accuracy, which remained at about 60% correct responses. As would be expected, fine-scale pixelation keeping both coarse-scale and relatively fine-scale information in the image yields better recognition in general compared to the intermediate-scale pixelation where some of the local cues of the source image have become already degraded by blocking the spatial contrast distribution. Similarly, it is natural to expect that by increasing image duration more information useful for veridical recognition can be integrated and accuracy increases. Indeed, with fine-scale depiction of image cues and features, a systematic increase in exposure duration up to 1000 ms monotonically increased recognition accuracy up to the level of perfect perceptual performance. Somewhat more interestingly, with intermediate pixelation coarseness only the initial increase in duration helps improve accuracy up to a moderate level of performance, but the perceptual system does not benefit any more with an additional duration increase (up to 1000 ms). Thus, with 20 ms duration most of what can be perceptually extracted from the pixelated image to reveal the category of the source image when it is spatially quantized at the "medium-coarse" level is available and accessible. However, the intriguing part of the results pertains to the experimental conditions when the coarse scale of pixelation was used. With 1 ms duration correct recognition for the 36^2 pix/im pictures was low, at about 30% correct. Similarly to the 56^2 pix/im condition duration increase up to 20 ms elevated accuracy to about 50% correct. Surprisingly, a further increase in image duration from 20 to 1000 ms *decreased* the level of correct recognition considerably—this was merely a 20% correct level of recognition. Thus, in addition to the expected result that the coarsest level of pixelation leads to the lowest level of recognition performance in general, there seems to be a curious nonmonotonic function of the level of correct perception as a function of exposure duration (with certain intermediate duration being optimal for best possible recognition). Importantly, by increasing exposure duration of the coarse-quantized image the accuracy of recognition of the source image decreases. How could these results be explained?

An extended excerpt from Sergent (1986) seems to be a pertinent theoretical background for attempting to explain these results: "The visual system owes its great sensitivity in large part to its capacity to integrate luminous energy over time. Visual acuity develops progressively, and fine details become discernible later as energy is summed sufficiently to resolve the higher acuity requirements for these details. At a psychological level, this results in a percept of gradually increasing clarity, starting with the perception of a diffuse, undifferentiated whole which then achieves figure-ground discrimination, and finally a greater distinctiveness of contour and inner content. At a physiological level, activation of the various visual channels varies as a function of their respective integration time, and low frequencies are integrated faster than high frequencies, making the content of a percept initially a function of the low spatial-frequency spectral components of the stimulus. Although it is difficult to realize and to be aware of this developing percept given the very brief time during which integration of luminous energy takes place..." (Sergent, 1986, p. 24). Sergent also posits that increased perceptual clarity and detail of the stimulus-image along its microgenetic development in terms of information content allows for new operations to be performed. Initially, general categorizations can be made, but no intricate identification within the same category. With additional perceptual tuning progressively at a finer spatial scale identification becomes possible, but it is based on local element configuration rather than very fine local feature detail. Only at the end-stages of percept development can local features also be individuated and clearly discriminated. Thus, for some tasks the microgenetic later stages are redundant (because already the earlier-stage representation allows for performing that task) but for some other tasks additional percept-development stages are necessary. It is obvious that the theoretical picture presented by Sergent (1986) essentially reminds the understanding of percept formation suggested by the microgenetic research from the late 19th and early 20th centuries (see Bachmann, 1980, 2000, for review). Importantly, this picture is also consistent with more recent psychophysical, psychophysiological, and computational-modeling research showing similar regularity of visual spatio-temporal integration (Baron & Westheimer, 1973; De Cesarei & Loftus, 2011; Gao & Bentin, 2011; Goffaux et al., 2011; Hegdé, 2008; Hughes et al., 1996; Loftus & Harley, 2004; Neri, 2011; Parker et al. 1996a,b; Vassilev & Stomonyakov, 1987; Watson, 1986; Watt, 1988). How does the SF anisochronic conceptualization of visual image microgenesis help to explain the results of Bachmann (1987)?

Essentially, the spatially quantized (pixelated) image, when depicted at intermediate or coarse levels of pixelation, represents inherently inconsistent images. In source images that are not spatially quantized and in images with very fine-scale spatial quantization (eg, more than about 100 pix/im for objects unidimensionally) there is no inconsistency between global configuration and local cues helpful for recognition or identification. However, in the more coarsely quantized images, coarse-scale global-holistic cues are helpful for verifical perceptual processing of the source-image content but the local-level cues represent blocks of the quantized image that are not helpful, may be misleading and can interfere with veridical perception by acting as masking information. The pixelation transform applied to generate spatially quantized (blocked) images always leaves more coarse-scale, holistic information of the source image less distorted compared to the fine-scale, local information. Because almost by definition the helpful information is spatiotemporally integrated earlier than the misleading or masking information, longer durations of a pixelated image must be bringing in more masking and/or attentionally misleading effects than shorter durations. With the latter, misleading/masking information is not sufficiently integrated into the evolving (forming) percept and cannot so easily obscure the veridical holistic coarser scale information. These considerations explain why with coarsely pixelated images the initial increase in duration increases recognition but the further increase of duration decreases it (Bachmann, 1987): initially the holistic coarse-scale information becomes better integrated with exposure duration being increased while local fine-scale masking/misleading information is still absent and only with an additional increase in duration does the masking/interfering effect accrue as long as the uninformative, conflicting detail is added to the microgenetically forming image.

Theoretically, the cause of impairing perception due to the blocking-transform of the original image must not be unitary. Several detrimental factors and impairment mechanisms can be listed. (1) *Loss of local fine-scale feature information* because of the local brightness averaging within the blocks which eliminates fine-scale local luminance contrast gradients and replaces them with a uniform and equiluminous areas. This cannot be the cause of the effect of decreasing performance with increasing the duration because availability of this information is zero independently of exposure duration. Furthermore, the fact that the initial increase in duration also improves coarse-scale quantized image recognition and the level of recognition is many times above the correct-by-chance level altogether show that stimuli could also be recognized moderately well according to their coarse-scale information. (2) *Masking of relevant low-SF information by irrelevant high-SF information* carried by the edges of the blocks, executed by interchannel inhibition. At first this explanation seems valid because the fine-scale edges as maskers become ever better represented with increased duration and their masking power increases. However, as explained earlier, research has shown that masking between SF channels separated by a large difference along the SF spectrum is not effective (Durgin & Proffitt, 1993; Morrone & Burr, 1997; Morrone et al., 1983). (3) *Higher-level masking* where the two object representations—coarse-scale source-image representation and coarse plus fine-scale mosaic representation—compete for the service of the higher-level interpretation mechanism. Initially microgenesis forms a coarse-scale representation of the source image, but gradually the fine-scale information is integrated with additional duration and an object that is easier to categorize as a mosaic of blocks than a source-image (eg, a face) becomes represented. Provided that the fast categorization has not been made before the formation of the mosaic image with its characteristic sharp edges and corners of variable luminance blocks, the newly emerged object masks the earlier-formed object. In this case it is not a between-channels sensory masking, but masking between different object interpretations where the following object replaces or substitutes the preceding object in explicit perceptual representation or working memory [compare also similar explanations used in the earlier mutual-masking research by Bachmann and Allik (1976) and Michaels and Turvey (1979), and microgenetic masking research by Calis, Sterenborg, and Maarse (1984), and Leeuwenberg, Mens, and Calis (1985)]. This explanation is consistent with the results (Bachmann, 1987) showing a decrease in recognition with an increase in duration. We must note that here we are dealing with a somewhat unusual variety of masking that is interesting in its own right. Traditionally and prevailingly, masking has been brought about and effected by two or more alternative stimuli. Here, one and the same visual stimulus includes mutually inconsistent aspects that unfold perceptually with different time courses. Thus, in essence, this is an instance of masking of the stimulus by itself, which was called "autoclitic masking" (Bachmann, 1987). Some aspects of the stimulus structure (eg, local detail or one category interpretation) mask other aspects (wholistic configuration and/or other category interpretation). (4) *Distortion of configuration.* More or less coarse-scale spatial quantization not only filters out certain fine-scale local information of the source image, but also creates a new stimulus-configuration. Local square-shaped blocks change the spatial arrangement of the brighter and darker areas that define luminance contrast spatial distribution of the original image. Compared to the original, the coarse-pixelated picture becomes more "edgy," rough-looking, with local brighter or darker areas twicthed or squeezed so as to become the local blocks of the image. As configuration is one of the prime cues for recognition, its distortion causes problems for easy recognition. Moreover, to a certain extent the blocking-transform, by means of creating a new configuration, may lead to replacement of the more veridical early configuration formed at the initial microgenetic stages by the new configuration formed later in the course of percept formation. This explanation is consistent with the experimental results (Bachmann, 1987).

There is an aspect to the results showing a nonmonotonic function of recognition of the coarse-quantized images which is theoretically especially significant. The fact that with a certain shorter duration veridical recognition is much higher than with a longer duration shows that spatial quantization has a masking or degrading effect not because of restricting the availability of useful information by some elimination or filtering procedure, but by influencing the processes that use and/or interpret information represented by earlier stages of processing. These processes can be not only the perceptual ones, but also of an attentional heritage. One possibility is that spatio-temporal attention is gradually tuned (zoomed) from coarse-scale image contents to fine-scale contents. The other possibility is that attention switches from the first-formed version of object representation to the later-formed object representation. This takes place within the processing domain where perceptual-cognitive wholistic configural objects compete for services provided by focal attention.

When studying the effects of temporal variables on pixelated image perception, Bachmann's (1987) work had several limitations. The range of durations and spatial quantization levels used was limited and the results were thus that rather qualitative, but not parametrically tuned quantitative, effects could be precisely established. Furthermore, when increasing stimulus duration the factors of duration and luminous energy are confounded. (Bloch's law, $I \times t = \text{const}$, states that with stimulus durations less than about 100 ms luminous intensity and duration have equivalent effects on perceived brightness.) This means that we cannot be sure whether primarily the integrated luminous energy was the prime causal factor in determining the effects or the duration per se was important in obtaining the effects. The recognition task used in the study was such that it can also be quite well performed based on the coarse-scale representation. But what about identification among the objects belonging to the same general category? Also, in the task used by Bachmann (1987), subjects did not know that specifically quantized images would be shown and they were not prepared for this class of image. What could happen when subjects know that images they have to use to complete the task are exclusively the pixelated type of images? In the Bachmann (1991) study, these problems were overcome in two experiments.

In Experiment 1, six alternative gray-level source images depicting frontoparallel views of male faces were used, each pixelated at eight levels of pixelation: 15, 18, 21, 24, 27, 32, 44, and 74 pix/f. Observers had to identify each pixelated picture presented with varying durations—1, 4, 8, 20, 40, and 100 ms. The increase in correct identification was steep with an increase in exposure duration from 1 to 4 ms (medium and fine level of pixelation) or 8 ms (coarse-scale pixelation at 15 pix/f) and slowed down thereafter. However, in Experiment 1 an increase in stimuli duration was confounded with an increase in integrated luminous energy because all values of duration were 100 ms or less, which is known to be the condition for temporal integration (according to Bloch's law, $I \times t = \text{const}$, satisfying the criterion of invariant perceived brightness). Therefore, we do not know whether the effect of duration on identification was a pure duration effect or an effect of integrated luminous energy. In order to control for this confound, in Experiment 2 luminance of the stimuli was adjusted to each duration value in a preliminary experiment so that apparent brightness of the stimulus display remained constant throughout Experiment 2. Also, a slightly different set of duration values was used, including one longer duration: 2, 4, 8, 100, and 200 ms. The effect of duration depended on the level of pixelation. With pixelation values from 18 to 74 pix/f, an abrupt increase in identification accuracy was obtained with an increase in duration from 2 to 4 ms (60–75% correct at 2 ms depending on pixelation coarseness and 80–90% correct at 4 ms). From 4 to 100 ms duration identification accuracy remained roughly the same and from 100 to 200 ms duration there was a slight further increase in accuracy by about 5%. However, with 15 pix/f a pattern of results similar to what was found by Bachmann (1987) was found: duration increase from 2 to 8 ms led to a steep increase in identification (from 46% correct to 69% correct), but an additional increase in duration from 8 to 200 ms caused accuracy to drop to 57%. This result was explained by an attention-dependent microgenetic process of percept formation proceeding from coarse-scale representation to fine-scale representation. Initially, an increase in duration benefits perceptual image buildup across the full-scale (broadband) spectrum of spatial contrast, but additional duration increase leads to a gradual perceptual tuning onto a progressively more fine-scale spatial representation. Because with coarse-quantized images the fine-scale contents provide masking noise of the blocks' edges, identification decreases. Attention has become tuned to (zoomed onto) the fine-scale misleading content and the coarse-scale more or less veridical cues of stimulus-configuration become unattended. The results obtained by Bachmann (1991) once more point to a seemingly counterintuitive effect—sometimes less time for an encounter with a stimulus is better than when more time is allowed for stimulus inspection.

The accuracy of perception of pixelated images as a combined effect of exposure duration and level of pixelation is illustrated in Fig. 6.1. However, it must be borne in mind that this kind of regularity is sensitive to the task, types of stimuli, and effective visual acuity.

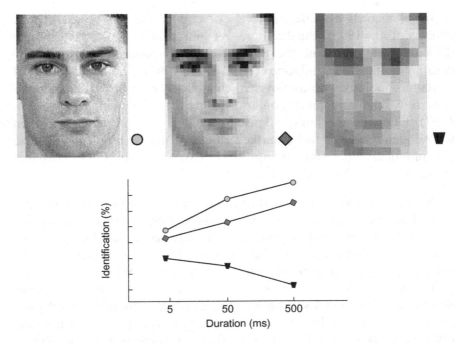

Figure 6.1 Accuracy of perception as a function of image duration and level of pixelation (a generalization from different experiments drawn for illustrative purposes). Upper curve: Fine-scale pixelation. Middle curve: Intermediate-scale pixelation. Lower-most curve: Coarse-scale pixelation. Identification of coarse-pix images can decrease with duration.

6.3 LIMITATIONS AND OPTIMIZATIONS IN THE APPLIED CONTEXT OF PIXELATED IMAGE PERCEPTION

Results from basic research on pixelated image perception include various regularities and effects that can be useful for practitioners to take on board. For instance, the effect of optimal exposure duration (ie, best performance identification or recognition with optimal image duration, Bachmann, 1987, 1991) can be used to maximize reliability of degraded image decoding, recognition or classification by system operators. On the other hand, a so-called spatial primary unit measure of a given form or pattern could be found to be a benchmark by finding the pixelation level with which an increase in exposure duration does not change identification or recognition accuracy over an extended range of durations. Finding this kind of level may be useful for optimizing the amount of information for image database storage capacity without compromising ease of recognizability independent of duration. Otherwise, the minimum image quality required for retrieval of facial records at different levels of confidence can be established (Bhatia et al., 1995). Furthermore, recognizability of pixelated images difficult to interpret in the applied contexts could be enhanced by combining image degradations (Uttal et al., 1997). The development of avatar technology with audiovisual capabilities may benefit from perti- nent research using pixelation transforms (eg, Campbell & Massaro, 1997). Knowing the limits of pixelation level beyond which a person whose identity needs to be masked can nevertheless be identified, also has practical value (Bachmann, 1991; Costen et al., 1994; Demanet et al., 2007; Lander et al., 2001).

Besides the above-mentioned research, there are studies more directly related to different kinds of applied tasks. For example, Thompson, Barnett, Humayun, and Dagnelie (2003) evaluated a model of simulated pixelized prosthetic vision for testing the effects of phosphene and the corresponding grid parameters on face recognition. With less than 70% grid random cell signal dropout, facial recognition was fairly good. Several gray levels and sufficient contrast are also advisable. Thompson and colleagues concluded that reliable face recognition with crude pixelized grids is possible, suggesting that quite coarse spatial resolution of prosthetic stimulation is tolerable.

The work of Pérez Fornos et al. (2005) also belongs to the paradigm of simulating artificial vision in order to determine the basic parameters for visual prostheses. One topic in such a domain is restoration of reading abilities. Stimuli information content can be reduced by prepro- cessing image with coarsening pixelation. Pérez Fornos et al. (2005) showed that real-time pixelation of text images that were presented for reading required about 30% less pixels than the offline versions. Square-shaped pixelation relatively favored real-time pixelation (with Gaussian pixelation as the control condition).

Scene categorization is a task encountered in many applied contexts. For economy of database and communication system capacity and speed of operation of the system it is desirable to use as small a number of bits as possible, but guarantee that performance in recognition, classification, or search would not be compromised. One of the central questions is this: how many pixels make a usable image or in other words—what is the minimal image resolution at which the visual system can extract the gist of an image of a scene or object? Torralba (2009) showed that very small thumbnail images provide sufficient amount of information to identify the semantic category of its gist when 32 × 32 color-level pixels define these images. Moreover, four to five objects from these scenes can be reported with 80% accuracy, although in isolation the objects with that level of resolution are unrecognizable.

Information processing and storage economy also dictate that it may be useful to create images that are easy to recognize, although a lot of information is eliminated and the gist or some other characteristic cues of the original image are abstracted. For example, Gerstner et al. (2012) developed an automatic method usable for producing abstract, low-resolution, well-recognizable images from high-resolution original images. This method had an advantage over the naïve nonautomated methods.

If large amounts of numerical data need to be communicated, graphical methods can be used allowing to capitalize on the capacity of the visual system to process information in parallel over an extended spatial area and use wholistic pattern/form perception routines. Pixel-based visualization also belongs here and is widely used in meteorology, remote sensing, bioinformatics, finance, etc. Temporal pixel-based visualization to simulate monthly temperature variation over an extended time period was used to assess perceptual proficiency across different tasks (Borgo et al., 2010). Surprisingly, pixel-block resolution had a relatively small impact on the effectiveness of visualization, but color-map variables mattered more. The most important effects stemmed from the perceptual task load.

One applied field where the effects of perception of pixelated images may have quite dramatic consequences concerns perperator or suspect identification from CCTV or other images of poor quality. We need to know what the levels of pixelation sufficient to trust eyewitness responses and choices are and how different factors influence this. Eyewitness evaluations can be made directly based on the image at stake or a degraded image can be compared to pictures of potential suspects available in databases or files of the case. Bindemann et al. (2013) studied how well observers can compare pixelated pictures with high-resolution photographs of suspects. Targets were unfamiliar persons. Performance dropped to chance level with 8 pix/f resolution. Reduction of the size of the pixelated image improved matching performance, because even with pixelation level at 20 pix/f accuracy remained at about 60% correct; relying on eyewitness reports must be taken with serious caution when evidential material consists of pixelated images.

The ever-present issue of finding a compromise between the requirement to minimize the number of bits communicated via the system on the one hand and to maximize user-friendliness and to guarantee avoidance of loss of necessary information on the other hand appears with obvious clarity in the domain of mobile TV and videophone development. In the experiments by Knoche, McCarthy, and Sasse (2005) one of the aims was to identify the minimum acceptable image resolution of mobile TV for a range of different bitrates and content types in relation to the impact of reduced image resolution. The acceptability measure was defined as the proportion of the subject sample that finds a given quality level acceptable all of the time during the videoclip viewing. The different contents were news, sports, music, and animation. Knoche et al. (2005) found that acceptability was significantly lower for image resolution smaller than 168×26 pixels universally for all contents. Viewers were least satisfied with image quality with sports clips and relatively most satisfied with animation. Sensitivity to the changes of the level of image resolution was higher when the bitrate was higher. Subsequently, Knoche et al. (2008) showed that with long shots the level of acceptable detail was 240×180 pixels.

CHAPTER 7

Combining Image Degradations

Pixelation transformation with progressively more coarse scales (ie, smaller number of pixels per image) is by no means the only or most essential image degradation. Images, including pixelated ones, can be degraded by adding random noise, filtering out certain spatial frequencies (SFs; in the Fourier domain or by Gaussian filters), decreasing the number of grayscale levels of contrast, etc. Importantly, if a pixelated image is formed or the original image is obtained or captured by some device, further degradations can be added to this pixelated image. It may be useful to know what effects other degradations besides pixelation can have on the perception of pixelated images.

In the seminal work by Harmon and Julesz (1973) pixelation by blocking was followed by SF filtering, leading to enhancement of image recognition. High-frequency SF components carrying the spurious noise of the edges and small corners of the image blocks were filtered out, and thus the coarse-scale gross configuration of the face could be more conspicuously experienced. (See additional discussion of this and related work in Section 3.1.)

Bhatia et al. (1995) employed a two-alternative forced-choice task with thousands of pairs of images (500 ms duration). Human faces as target images were paired with nontarget images (nonhuman heads, scrambled images constructed from human faces, inanimate objects). In addition to five levels of pixelation (from 8×8 pixels/face up to 128×128 pixels/face) gray-level scales over which the pixelated images were shown also varied between 2, 4, 8, and 16 gray level values used. The condition with only 2 gray levels of the 8×8 pixels/face images produced poor results—less than 20% above-chance correct responses. But the 16×16 pixels/face and more fine-scale images did not suffer much from being depicted by only two levels of luminance contrast (all discriminated at more than 90% correct). The results showed that using more than 8 gray levels does not give a further increase in correctness of perception of the pixelated images. A probit analysis

Perception of Pixelated Images. DOI: http://dx.doi.org/10.1016/B978-0-12-809311-5.00007-6

revealed that the more grayscale levels were used, the smaller the number of pixels per image necessary to reach the 75% correct response level. The 75% correct response threshold dropped from about 11 pixels/face when only one gradation of gray level was used to about 6.5 pixels/face with 8 gray levels.

Restoration of the pixelated image recognizability by SF filtering by Harmon and Julesz (1973) was demonstrated with relatively large stimulus images, but it was not clear from these results whether the same effect could also be found with smaller images. When Uttal et al. (1997) used large faces (about 6 degrees vertically, presented for 100 ms), coarse-scale pixelated images (about 18 pixels/face vertically) were perceived better in a 12-alternative identification task when SF-filtered, compared to the unfiltered control condition. This result replicated the original findings by Harmon and Julesz (1973). However, when the low-pass SF filter with the most coarse value was used, instead of an improvement in accuracy, faces were identified less successfully compared to the control condition without SF filtering. Thus, the effect of SF filtering of the coarse-scale pixelated images may be an improvement of perception as well as impairment, depending on the cutoff value of the SF filter. (When Uttal et al. (1997) applied SF filtering on the original, untransformed faces and only thereafter applied coarse-scale pixelation by blocking, impairment of perception was found.) Even more intriguing results were obtained when Uttal et al. (1997) used small facial images subtending about 6 degrees vertically. In the pixelation-only control condition percent correct responses with 30 vertically measured pixels per face stimuli amounted to more than 90% while decreasing the number of pixels to 12 lead to only about 35% correct responses. Curiously, when low-pass SF filtering was applied to 30 pixels/face, identification accuracy decreased, but in the 15 and 12 pixels/face condition low-pass SF filtering applied on the pixelated images helped increase accuracy. These results casts doubt on the generality of the Harmon and Julesz (1973) original results and show that the possibility of reinstating pixelated image identifiability depends on a combination of the size of the image, coarseness of pixelation, and cutoff value of the SF filter applied on the pixelated image contents.

Not only facial images have been subjected to multiple, combined image degradations to examine their effects. Twelve solid silhouettes of aircraft subtending about 1 degree of visual angle were used as target stimuli by Uttal et al. (1995c). In a discrimination task, observers were asked to specify whether the two sequentially presented stimuli separated by 1 s and presented for 100 ms were the same or different. Three different types of stimulus degradation were used in various combinations—random punctate visual noise added to the image area (ranging from 20% to 80%), low-pass SF filtering applied, and pixelation by local area brightness averaging executed. When relatively high levels of noise from 50% to 80% were added, discrimination suffered the most. The most important result of this study was that the order with which different types of other degradations were carried out on stimulus images had only a minor effect on discriminability, if any. While earlier work on face recognition had shown that blocking followed by low-pass SF filtering had different effects compared with the opposite order of degradations, with small nonfacial silhouette stimuli presented sequentially in the task of discrimination degradation, order did not matter to any substantial degree. This implies discriminative commutativity of the orders of pixelation and SF filtering degradations. Combination of low-pass filtering and visual interference by noise had a powerful detrimental effect on discrimination compared to the conditions where either degradation was used alone; thus, no demasking could be observed. A similar result was obtained when combining pixelation and visual noise. Low-pass filtering when added to pixelation had a negligible effect compared to the pixelation alone condition, except when the lowest value SF filter of 0.44 cycles/degree was used. With a reverse order—pixelation added to low-pass filtering—similar results were found. The blocking added little to the effect of low-pass filtering. Importantly, discrimination scores for blocking-to-filtering and filtering-to-blocking conditions were virtually identical. An important hint as to why pixelation and SF filtering demonstrate commutativity of order when combined and when assessed by discrimination performance comes from Figure 10.1 of Uttal et al. (1995c). The shape primitives of the solid silhouette images have very similar appearance with either order of degradations having been applied. However, internal configuration processing based on gray-level gradients within a facial or other images where internal element configuration plays an important role must be more sensitive to the order of these two degradations because when pixelation by blocking is applied after SF filtering, configuration is distorted more than when the opposite order of degradations is used.

Uttal et al. (1995c) also stress that discrimination and recognition are not basically similar as operations of visual information processing in terms of information these operations utilize. Recognition may primarily depend on global, low-SF information, but discrimination on local, high-SF information.

The effects of combining image degradations in an identification task were studied by Uttal and colleagues in a follow-up paper (Uttal et al. 1995a). Again, silhouettes of aircraft were used as stimuli (100 ms), but forced-choice 12-alternative identification was the task. Low-pass SF filtering, pixelation by blocking and 10% or 20% random dotted visual noise were applied as degradations. Visual noise universally caused a decline in identification performance. Surprisingly, when both pixelation and SF filtering were used, either order of imposing these two degradations produced improvement of identification compared to the single degradation condition. With filtering following blocking the result is predictable based on earlier results by Harmon and Julesz (1973) and other following studies. However, improvement with blocking following SF filtering (which happened with the coarsest scale pixelation and which produced a clearly perceptible pixelated shape with spurious gray-level variability between pixels) was surprising. The parsimoneous explanation for this effect is as follows. By the blocking transform different types of distinguishable, characteristic images with gray-level variability between pixels and with rectangular-cornered local features are produced. The prepixelated, filtered images of small silhouettes differ between each other only very slightly, but the images that are additionally pixelated are easier to distinguish because the coarse-scale blocks amplify the small differences between individual local image characteristics at the edges of the shapes.

Summarizing the findings of their three studies with combined degradations, Uttal et al. (1997) make several conclusions. First, the ways by which combined degradations affect stimuli perception depend considerably on the task, specifically on whether discrimination or identification (recognition) is required from observers. It is possible that these tasks depend partly on the different stimuli attributes they use and the expression of these attributes in turn depends on the particular combination and order of degradations. Second, the effects depend on the size of the stimuli, with small images helping to produce unexpected results, in that paradoxical enhancement may be achieved by either order of degradations. There was some indication that the effect of size may be independent of the SF which defines the stimuli. Uttal et al. (1997) suggest that instead of the distribution and interaction of SFs in the operation of image recognition, global or configural aspects of the spatial arrangement of the parts of a stimulus image as the Gestalt organization attributes are what steer the process to its result. Third, random-noise visual interference often inhibits recognition or discrimination (Uttal et al., 1995a, 1995c), but may also enhance correctness of perception (eg, Morrone et al., 1983).

Besides faces and aircraft silhouettes, more complex scenes have also been used as the stimuli for which the effects of different degrading manipulations are applied. In a carefully planned experimental investigation Torralba (2009) collected 240 images representing 12 scene categories such as street views, airports, forest, interiors, etc. He used six levels of image pixelation: 4×4, 8×8, 16×16, 32×32, 64×64, and 128×128. This was in order to downsample image information by a well-quantifiable way. Thereafter, these images were upsampled to 256 pixels and these images were shown to observers in the scene recognition task. In the 12-alternative forced-choice procedure images had to be categorized between 12 possible scene categories. Color images and grayscale images were both used. The image was displayed until the subject responded. Percent correct categorization increased in a virtually monotonic manner with an increase in the number of pixels. With 4×4-pixel images categorization was close to random guessing level (less than 20%). With 64×64-pixel images the correct response rate exceeded 90%. Color images were easier to categorize than grayscale images by about 10−20%. Color advantage was especially conspicuous with outdoor scenes compared to indoor scenes and the general level of correct categorization was lowest with indoor scenes. Thus, applying color-to-grayscale operation tends to decrease recognizability. Torralba (2009) posits that a scene pixelated at 32×32 pixels/ image is in most cases sufficient for correct categorization. The gist of a scene can be comprehended from coarse-level degraded images. Surprisingly, observers are capable of correctly reporting about 80% of the individual objects contained in the scene despite the fact that in isolation these objects are unrecognizable. Correct interpretation of a scene according to its low-resolution image also aids perceptual inferences as for the category of the constituent objects.

CHAPTER 8

Perceptual Task Effects

Several researchers of perception of pixelated images have drawn our attention to the dependence of the pixelation effects on the observers' task. (William Uttal and Denis Parker should be mentioned among the first in this context.) In the following paragraphs the effects of pixelation will be discussed specifically from this perspective.

8.1 IDENTIFICATION, RECOGNITION, AND DISCRIMINATION

The issue of task demands as related to spatial frequency (SF) processing in complex stimuli was raised by Sergent (1986). She particularly mentioned that identification and matching tasks may be different in terms of the key SF content necessarily used for each task. Identification tends to utilize relatively higher frequencies, while matching may be carried out as based on different scales of SF content. Thus, coarse-scale pixelation should be more complicated for identification than matching tasks. However, in both these tasks observers must rely on some perceptual operation of comparison between the stimuli with the set of alternatives more or less specified. In comparison with matching and identification, a recognition task when subjects have no explicitly available alternatives should prove to be most difficult. How do the results of different studies appear when analyzed from this angle?

A recognition task was used by Bachmann (1987) so that observers did not know the categories and identities of objects that may be shown. The level of recognition accuracy varied between 20% and 60%, when 36×36 and 56×56 pix/image stimuli were used. Thus, despite the moderate level of the scale of pixelation, performance was relatively poor. In a six-alternative identification task with the stimuli belonging to the same category (faces) Bachmann (1991) achieved a much higher accuracy level with this range of pixelation—between 70% and 90%. Only when the pixelation level was less than 18 pix/image did accuracy drop to about 50% correct. Six images of faces were used also by

Perception of Pixelated Images. DOI: http://dx.doi.org/10.1016/B978-0-12-809311-5.00008-8

Costen et al. (1994), pixelated at 11, 21, and 42 pix/image and the task was also that of identification. Again, with a known set of alternatives and the task of identification the accuracy level was quite high, ranging between 88% and 98% when coarseness of pixelation decreased. However, when in the second experiment Costen, Shepherd, Ellis, and Craw (1994) purposely selected the original faces that were more difficult to discriminate and included more pixelation values, the accuracy level of identification dropped considerably, ranging between 25% and less than 50% with 9 and 12 pix/image stimuli and reaching up to 70% and more with 23 and 45 pix/image stimuli. Thus, the effects of task interact with difficulty of discrimination between the alternative stimuli. With a similar set of stimuli and the same task, Costen et al. (1996) obtained essentially the same level of accuracy in their follow-up study—percent correct varying between about 30% and 60%.

Recognition without preknowledge was used by Lander et al. (2001). Faces of 30 famous and 10 noncelebrity faces were pixelated at approximately 10 and 20 pix/image levels and presented for 2.5 s. Accuracy of recognition varied between 46% and 53% for 10 pix/f images and between 67% and 76% for 20 pix/f images, depending on the viewing distance (either 1.5 or 3 m). A slightly higher accuracy level compared to the earlier recognition-task results (Bachmann, 1987) may be caused by the fact that Lander et al. (2001) used stimuli belonging to the same category and because celebrity faces are overlearned by representatives of subject population. Thirty-six celebrity facial images were used in a similar task also by Sinha et al. (2006). Either full-face degraded images or degraded images depicting only the internal part of faces were presented for recognition. Presentations started with the coarsest pixelation level and progressed toward a systematically finer level of pixelation. Increasing the number of pixels per image from 5×7 up to 150×210 pix monotonically increased recognition accuracy of full-face images from about 30% to 100% correct. The pixelation levels typical for the earlier studies by Bachmann and Parker (eg, 9×13 and 12×17 pixels in Sinha et al.) produced accuracy levels at about 70–90%. Sinha and colleagues suggest that overall head configuration and characteristic cues of celebrities may help reach quite high recognition levels. However, when only internal parts of faces were used, the recognition level was decreased dramatically. With 12×17 pixels and coarser, recognition was essentially random. With 19×27 pixels per image, recognition exceeded 30% and with 37×52 pixels it was approximately 70%. The effects of pixelation on recognition and identification depend considerably on the number and preknowledge of alternatives, whether full-image or only internal parts of image configuration are used, and how similar or dissimilar the alternatives are. Also, effects of learning play their part.

The task used by Bhatia et al. (1995) was different. They used two-alternative forced choices, where observers had to decide whether the human face was presented on the left or right in the pairs of pixelated stimuli (shown for 500 ms). Thus, the direct comparison of the image cues was possible to aid discrimination and recognition. As could be expected, performing this task produced a considerably higher level of accuracy and not only due to the high level of correct responding by chance (50%). With 16×16 pix/image and finer scales accuracy was between 90% and 100%, but even with 8×8 pix/image accuracy varied between 70% and 85%, depending on the number of gray levels used.

In a somewhat similar discrimination task, Uttal et al. (1995c) asked observers to decide whether the two sequentially presented pixelized stimuli were the same or different (again, 50% correct-by-chance, level). Aircraft silhouettes were used as the stimuli. With this relatively simple task at hand pixelized images were discriminated with almost perfect accuracy. When Uttal and co-workers used an identification task (Uttal et al., 1995a), accuracy considerably depended on level of pixelation, ranging from almost perfect identification with moderate number of blocks per image to less than 70% correct with the coarsest pixelation at 2.6 pix/image. In a follow-up study, 12 alternative faces were the stimuli, pixelized at four levels (approximately 60, 35, 24, or 18 pix/image). When the stimuli were large (6 degrees along the vertical dimension), identification accuracy was surprisingly high despite degradation brought about by pixelation: near 100% with the two finest pixelated images, almost 90% correct with 24 pix/f and about 80% correct with 18 pix/f images. However, as the most coarse pixelation level (18 pix/f) was still finer than the most coarse pixelation levels in Bachmann (1991) and Costen et al. (1994), the discrepancy of the accuracy levels between Uttal's and other work can be explained as a result of the difference between the limiting values of pixelation coarseness. Moreover, in the Uttal et al. (1997) experiment observers had preliminarily overlearned the original images of faces, which must have increased their identification skills. When the stimuli were small (1 degree vertically), identification accuracy strongly depended on the scale of pixelation. With 30 pix/f, accuracy exceeded 90%, with 20 pix/f it exceeded 80%, but already with 15 pix/f accuracy dropped below 60%. With 12 and 10 pix/f pixelation levels an accuracy level of about 30% correct responses was low indeed. Thus, the task effects in their interaction with level of pixelation also depend on other variables such as the size of stimuli and perceptual skills.

A six-alternative identification task was used also by Morrone and Burr (1997). Coarse quantized images of faces with 12×12 pixels were prepared for 1000 ms presentation. With supraoptimal root mean square (RMS) contrast levels, identification accuracy exceeded 50%, with the majority of results appearing between 60% and 80% correct responding level.

Pixel-based visualization techniques have become a widely used IT approach. Borgo et al. (2010) studied the effects of block resolution and task on the capacity to read information correctly. Surprisingly, block resolution had only a limited impact on the effectiveness of pixel-based visualization, but the tasks with different load had a clear effect. (Level of accuracy changed from 40% to 75% when high-load tasks were replaced by low-load tasks.) Importantly, as variable color pixels were used, Borgo et al. (2010) showed that the color palette played a central role in predicting the efficiency of pixel-based visualization.

8.2 SPATIAL PRECUEING OF COVERT ATTENTION

In the majority of tasks where pixelated images are to be perceived, attention is focused on the spatial location from where the image is presented. Attentional competition is not involved. Invariance of the experimental conditions in terms of spatial attention orienting and selection of attentional focus has been typical for pixelation studies. An exception to this rule was presented by Bachmann and Kahusk (1997) when they examined the effects of spatial attention precueing on the perception of pixelated images. Several earlier works had suggested that not only low-level sensory processing mechanisms tuned to differ-ent SF scales of images are involved in pixelated image perception, but also more high-level cognitive mechanisms related to perceptual organization and mechanisms of spatial attention may be implicated by the experimental results (Bachmann, 1991; Morrone et al., 1983; Uttal et al., 1995a, 1995c). In two experiments, Bachmann and Kahusk (1997) presented pixelated images of faces and unpixelated original con-trol stimuli at one of four possible locations around the fixation point with excentricity equal to 2 degrees. Pixelation levels varied between 9 and 16 pix/f; stimuli duration varied between 28 and 76 ms. Stimulus alternatives were presented unpredictably at one of four locations either without precueing or were precued by a 20-ms local precue that was presented before the facial stimulus with stimulus onset asynchrony (SOA) = 120 ms, indicating the position where the following face image was presented. An eight-alternative forced choices identification task was used for examining the effect of precueing on target perception.

In Experiment 1, precued and noncued original faces were perceived at an equally high level of accuracy close to 80% correct. When face images were pixelated at four levels between 12 and 16 pix/f, precueing significantly helped improve correct identification compared to the non-cued trials (about 60% correct vs 50% correct, respectively). However, contrary to what one would expect from typical effects of spatial-attentional precueing, precued trials lead to lower-level identification accuracy compared to noncued trials when pixelation levels between 9 and 11 pix/f were used (percentages of correct identifications less than about 40% and almost 50%, respectively). In Experiment 2 luminance of the precues was decreased in order to avoid possible forward masking effects of precues on target image perception. Also, the size of the precue was made equal to the size of the target images of faces. Moreover, a smaller value of facial image duration was added (4 ms) as well as a longer duration (512 ms) added in order to test the possible effects of target brevity and eye movements on task performance. (With 120 ms SOA between precue and target and short target durations there is not enough time for effective eye movements for fixating the target area and any attempts to move fixation from the fixation dot to target area typically soon show to the observers that this strategy is counter-productive, leading to a decrease in distinct vision of the targets. Thus true covert spatial-attention orienting is guaranteed. However, with 512 ms target duration there is enough time to move the eyes to fixate on the target spatial area and increase effective acuity.) Experiment 2 produced the following results, replicating the most important aspects of the results of Experiment 1, with minor additional new data. Most likely due to the addition of the very short target duration value and due to the change of precue spatial size and contrast, precueing now improved identification in the control condition with unpixelated target images (almost 80% correct vs 60% correct in the noncued condition). However, with coarsest levels of pixelation (9–11 pix/f) precueing again produced lower-level identification (above 25% correct) compared to uncued trials (about 35% correct). These data were obtained with all target durations except 512 ms, the trials of which were excluded from this analysis in order to examine results exclusively for the conditions of covert orienting without effective eye movements. When data from trials with long target duration were analyzed, no effect of precueing whatsoever was observed: 90% correct for control condition (both pre-cued and uncued trials), slightly less than 60% correct for 12−16 pix/f stimuli (both precued and uncued trials), and about 40% correct for 9−11 pix/f stimuli (again, both precued and uncued trials).

The experimental results (Bachmann & Kahusk, 1997) suggested that precueing could pretune perceptual filters before actual signals from target images are input and this pretuning has its effect depending on the level of stimulus pixelation. The pretuning operation is hypothe-sized to act according to the coarse-to-fine zooming routine with regard to the spatial levels of image content (Eriksen & StJames, 1986; Watt, 1988). With intermediate levels of pixelation (eg, 12−16 pix/f) or unpixelated low-intensity images pretuning of spatial filters ignited by the precue has zoomed onto fine spatial scales within the 120 ms cue-to-target SOA and as soon as the target-image spatial signals arrive, filters are ready to process the fine-scale image information. With original images and intermediate-level pixelated images, finer scales are instrumental for identification as these spatial levels contain information useful for extracting the identity cues. Accuracy increases due to precueing because in the noncued trials the target stimulus itself drives the filter zooming and because this takes time and the durations of the stimuli are short, zooming level of the filter is suboptimal before the target input is switched off in these uncued trials. With coarse quantized images the effect of precueing means that for the time when target-image signals arrive, filters have been zoomed already at the fine-scale level and in this case this level represents meaningless blocks defined by their sharp edges and small corners of the blocks. Precueing, by virtue of presetting the zooming of filters according to the coarse-to-fine routine, helps to emphasize the misleading, masking image cues instead of the veridical configural information. Bachmann and Kahusk (1997) contended that precues serve the function of saving the seem-ingly "perception-empty" time of presetting the spatial filters, whose value of zooming parameters becomes optimized for the moment when target-image signals arrive. Coarse-scale filters have been already switched off and predominantly fine-scale representation of the stimulus will be utilized in perception. The accuracy of perception now depends on whether the fine-scale content is helpful for target identification or not. This in turn depends on the coarseness level of pixelation.

The somewhat counterintuitive effects of spatial precueing show that coarse-scale pixelated images represent an unusual type of perceptual stimuli for which some of the standard regularities of perception do not apply. Coarse-scale pixelated images can be considered as ambiguous or multistable perceptual objects, allowing alternative interpretation—a face (or half-tone image of a scene) on the one hand, but at the same time the same image can be categorized or interpreted as an abstract mosaic of blocks or isoluminant squares. When by some means such as longer duration or attentional zooming the emphasis is especially put on the cues aiding interpretation of the object as a mosaic instead of the original prequantization content, apparently paradoxical effects occur.

8.3 PERCEIVING EMOTIONAL EXPRESSION, BEAUTY, AND PERSONALITY

The effect of pixelation on the perception of object identity has been by far the most studied aspect in the paradigm of spatially quantized image processing. But it is also essential to know how the more socially contextuated characteristics of objects can be processed by the perceptual system when the images of these objects are pixelated. Some experimental research has payed attention to this and I will summarize the main findings from this perspective.

Emotion recognition from photographic images with deteriorated pictorial quality was investigated by Wallbott (1991). He showed that recognition of facial emotional expressions was not significantly affected by variations in picture size or contrast, but was affected by reduction of spatial resolution of the images by the pixelation transform. Photographs presented to observers for 10 s each and depicting seven basic primary emotions (Ekman & Friesen, 1976) were pixelated at three levels: 150×150, 75×75, and 38×38 pixels. (The obviously finer scale of pixelation compared to the studies of identity discrimination or recognition is understandable because visual cues to identity tolerate much more severe distortions than relatively more subtle cues of emotional expression varying in faces with invariant identity.) The average rate of emotion recognition for undistorted pictures was 0.64, for 150^2 pixels condition—0.59, for 75^2 pixels condition—0.56, and for

38^2 pixels condition—0.44. Considering that the chance expectancy of correct recognition is about 0.14 it is clear that pixelation does not eliminate the emotion recognition capacity and that only with relatively coarse level of pixelation (38^2) there is some notable drop in sensitivity to emotional categories of facial expression. Combining the coarsest pixelation with lowest contrast of the image, the recognition rate dropped to 0.26, which is still definitely higher than chance level. (Replicating the typical earlier findings, different emotions were recognized overall with different accuracy, with happiness at 0.97, anger and surprise 0.63, sadness 0.59, fear 0.51, and disgust 0.44.) The tolerance of emotion recognition to spatial deterioration of facial images shows that the system of communicating emotion must be robust and suitable for social communication at medium-range distance.

Compared to what Wallbott (1991) had found, Japanese researchers Endo, Kirita, and Abe (1995) showed even much higher sensitivity to emotion expression was communicated by pixelated images. They presented 4.2 degrees wide images of female faces depicting Japanese models for 2 s, with spatial resolution varying in different experiments between approximately 23, 11, 8, 6, and 4 pix/f along the horizontal dimension. Three emotion categories were used: happy, sad, and neutral. The task was a forced-choice binary decision—whether the face was happy or neutral in one condition and whether the face was sad or neutral in the other condition. Happy expressions from pixelated pictures were discriminated with almost the same accuracy as from the original pictures across all coarse pixelation levels, except for a small drop in accuracy with the coarsest quantized pictures (eg, 6 pix/f). Sad expressions became systematically more difficult to discriminate when pixelation levels were coarsened from 11 to 8 pix/f and finally to 6 pix/f. Endo et al. (1995) noted that the perception of happy expression is mediated more by low SF (LSF) information while the perception of sad expression was more by higher SF (HSF) information. [A similar LSF dependence was found for fearful expressions by Vuilleumier et al. (2003).] Additionally, reaction times to happy expressions were not slowed by pixelation, but sad expressions took longer to discriminate when pictures were pixelated.

It is also of interest what the image contents used for expression perception are compared to identity perception. White and Li (2006) hypothesized that the former aspect relies more on edge-based image cues (thus, HSF content) while the latter aspect more on configural cues (thus, LSF content). They presented simultaneous pairs of female faces (height 5.3 degrees) for 4 s and asked observers to decide whether the faces had the same expression (independent of facial identity) or whether the faces belonged to the same person (independent of facial expression). In this matching task, original faces, pixelated faces (20 pix/f horizontally), or blurred faces (Gaussian filter with 6-pix radius) were shown, depicting one of the four emotions—happy, angry, sad, and afraid. Both pixelation and blur slowed emotion discrimination compared to untransformed pictures. Matching the identities was equally fast with all types of images. The hypothesis about the reliance of emotion perception more on HSF content compared to identity perception received support. [When instead of the pixelation transform Aguado, Serrano-Pedraza, Rodriguez, and Román (2010) used low-pass and high-pass SF filtering, they did not find strong effects on error rate in gender and expression perception, but speed of responding was faster with HSF images compared to LSF images. Their filter values corresponded roughly to 14 pix/f in the LSF condition and 20 pix/f in the HSF condition.]

Similarly to facial-expressive cues, personal facial attractiveness is also an important cue in social communication and its effects depending on image pixelation must be studied as well. Unfortunately, there has been only one paper devoted to this topic (Bachmann, 2007). The experiment was set up as follows. In the preliminary stage of the study, photographs of 14 female faces were evaluated for their attractiveness on a 10-point rating scale and two sets of photos were prepared—highly attractive with average ratings above 5.5 points (five faces) and relatively unattractive with average ratings less than 5.5 points (nine faces). Then all pictures were pixelated at 10, 12, 14, 17, 20, 25, or 40 pix/f measured horizontally along the interauricular axis. Pixelated images were presented for 1.5 s and observers belonging to a new group of experiment participants were asked to evaluate subjective attractiveness on a similar 10-point rating scale. Total means of attractiveness evaluations with different levels of coarseness of pixelation ranged between 3.71 (in case of 10 pix/f images) and 4.84 (in case of 20 pix/f). The increase in the perceived attractiveness with increasing number of pixels per image (ie, increasing detail of the image contents) was due to the increasing attractiveness ratings given to the attractive face set, but the ratings for the relatively unattractive set remained roughly at the same level with varying coarseness of pixelation. Coarse quantized attractive faces appear much less attractive than fine-scale quantized and original attractive pictures, but for unattractive pictures an increase in the image detail did not add much perceived attractiveness. The average ratings for the set of unattractive faces in the most detailed depiction condition (ie, 40 pix/f) equaled 1.94 and for the set of attractive faces this estimate was 8.69.

Results showed a gradual, systematic increase in the attractiveness ratings with a decrease in the coarseness of pixelation; there was no "quantal leap" or abrupt step-like increase with certain specific value of pixelation. The 10 and 12 pix/f pixelation levels led to virtually the same level of attractiveness ratings of the two sets of facial images (attractive and unattractive), but the 17 pix/f level of spatial quantization clearly distinguished attractive faces from unattractive ones. Most of the pictures ultimately evaluated as attractive (both by the expert evaluators at the preliminary stage and by the observers in the main experiment) became rated as definitely more attractive by the 17 pix/f condition. Thus, by moving from 12 to 17 pix/f image quality, sets of attractive and unattractive faces become differentiated. Interestingly, the value of pixelation defining this step corresponds to the value of pixelation that was found to lead to an abrupt increase in identification of the personal identity of faces in earlier studies (Bachmann, 1991; Costen et al., 1994). It may be that attractiveness cues in facial images are somehow associated with identity cues. This relation, if any, and its putative foundations, should be studied in future research. The main message from the study (Bachmann, 2007) is that basic cues for facial attractiveness are communicated by coarse-scale information implicit in the facial image. This in turn means that configural cues seem to be the leading source of attractiveness evaluations.

Recent research has convincingly shown that people automatically produce appearance-based inferences about the traits and dispositions of the people they encounter (Todorov, Said, & Verosky, 2011; Zebrowitz, 2011). This happens also in zero-acquaintance situations and the process of trait inference takes only a fraction of a second. Although real personality characteristics are difficult to perceive veridically, social stereotypes in person evaluation are common and this is largely based on the overgeneralization of the cues that can be well perceived objectively, such as emotional expressions, age/maturity, gender, symmetry and attractiveness, etc. (Zebrowitz, 2011). In most of the research on face-based trait perception parametrical variations of the availability of image cues has been controlled either by temporal or spatiotemporal factors (exposure duration, masking) or by the target image transformations. However, up to now spatial quantization has been used for this purpose only occasionally. In one such study (Nurmoja et al., 2012) trait perception from pixelated images of faces was explored for personality dispositions and traits such as assumed criminality, trustworthiness, and suggestibility. Pixelation levels were 10, 12, 14, 17, 20, 25, and 40 pix/f in face width and exposure duration was 2 s. In the preliminary study a large pool of male and female faces was evaluated for the above-mentioned perceived-personality attributes and 10 faces that received the highest and 10 faces that received the lowest average ratings formed the two sets of images within each of the three categories (criminality, trustworthiness, and suggestibility). On average, female faces were treated as less "criminal," more trustworthy, and more suggestible compared to male faces. Already with 10 pix/f images high-criminality and low-criminality sets of faces were slightly, but significantly, differentiated by ratings. Beginning with 10 pix/f spatial quantization, ratings of low-criminality-perceived faces remained at a virtually similar level with an increase in the number of pixels per face, while ratings of high-criminality-perceived faces steadily increased. Trustworthiness versus untrustworthiness could not be differentiated with 10 pix/f images, but beginning with 12 pix/f pixelation value, high trustworthiness sets of faces produced systematically higher ratings of trustworthiness with progressively finer scale value of pixelation. Low-trustworthy-perceived faces kept roughly the same low level of trustworthiness ratings with an increasing number of pixels per face. At 10 pix/f quantization level highly suggestible perceived and low suggestible perceived sets could not be differentiated, but already with 12 pix/f the two sets had slightly different levels of rating. High suggestibility set ratings increased progressively with the number of pixels per face being increased. Low suggestibility sets of faces also slightly increased their suggestibility ratings with a decrease in the coarseness of pixelation. It seems that coarse-scale pixelated images of faces appear universally to be not quite suggestible. The main result of this study shows that subjective perceived dispositions and traits such as assumed criminality, trustworthiness, and suggestibility can be picked up from the quite coarsely pixelated images at 12 pix/f and up. Gross configural cues of facial appearance seem to be instrumental in making subjective perceptual inferences about human traits. (Fig. 8.1 illustrates some of the effects of pixelation level on face-based impression of personality.)

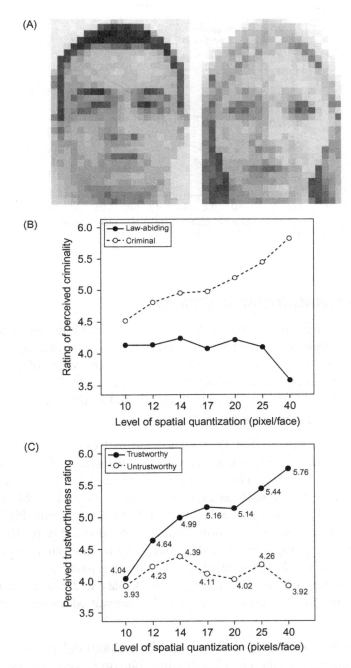

Figure 8.1 Facial-appearance-based subjective perceptual evaluations of personality as a function of the level of face image pixelation. (A) Examples of pixelated images. Ratings of perceived criminality (B), perceived trustworthiness (C), and perceived suggestibility (D). [Adapted from Nurmoja et al. (2012).]

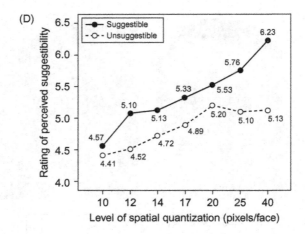

Figure 8.1 Continued

8.4 PERCEIVING DYNAMIC STIMULI

Pixelation transform can be used not only with static images, but also when the original image to be transformed is dynamic, that is, it changes over time, such as when emotional expression of a face is changing or visemes of a talking face unfold in time. There are but a few examples of research on the effects of spatial quantization on the perception of dynamic images and vice versa.

It has been shown that face identification from degraded images improves when dynamic images are presented instead of static images (Lander et al., 2001). These researchers used 2.5-s clips involving nonrigid movements, such as expressive dynamics and visible speech with some rigid motion of the head possible to some extent. Pixelation levels were 10 or 20 pix/f. Identification increased roughly by 10–15%. A clear practical implication of this study tells us that when the capacity of pixelation to hide the identity of a person is assessed with a static image, but then communication systems display a dynamic image of the same person, unexpected and probably unwanted recognition may occur.

Although Ehrlich, Schiano, and Sheridan (2000) did not use pixelation for image degradation (edge-finding, blurring, contrast, and some other transforms were used), they also found a beneficial effect of image dynamics compared to static image. Six basic emotions evolving from neutral expression up to a maximal level of emotion over 8 s were acted by a trained actor. Compared to the condition with normal image, degradation considerably impaired emotion ratings when static image depicting the endstage of the unfolding emotion was used. However, with the dynamic image, degradation did not change emotion ratings.

8.5 PERCEIVING VISIBLE SPEECH

The effect of spatial quantization on visual speech perception—that is, lipreading—was investigated by Brooke and Templeton (1990). The mouth region of a face articulating English vowels was shown, with pixelation level varying between 8×8 and 128×128 pix/f. Successful lipreading decreased when the pixelation level was reduced beyond 32×32 pix/f resolution.

Reading speech from a talker's face was the agenda also in the experiments done by Campbell and Massaro (1997). Stimulus items in visible speech recognition consisted of nine syllables differing in their initial consonant (/ba/, /va/, /tha/, /da/, /ra/, /la/, /za/, /ja/, /wa/). Dynamic facial images were presented as animations of the computer-generated synthetic heads (with about 30 frames/s). Vertical extent of the head image was about 17 degrees. Spatial quantization was achieved by a mosaic transform producing five levels of pixelation coarseness: about 650 pix/face along the horizontal dimension (control condition), 62, 32, 16, and 8 pix/f. Only the lower half of the face extending from the tip of the nose to the bottom of the chin was pixelated, with the rest of the face left unpixelated. Observers were instructed to respond by using one of the nine keys of the keyboard, each corresponding to one of the visemes which had to be identified. Identification accuracy dropped systematically with increase in coarseness of pixelation: from roughly 65% correct in the control condition to about 60% with 62 pix/f, 40% with 32 pix/f, 30% with 16 pix/f, and less than 20% with 8 pix/f levels. Visemes /ba/, /va/, /tha/, /la/, /ja/, and /wa/ were better identified than visemes /da/, /za/, and /ra/. In general, up to about 32 pix/f level of pixelation, visemes resisted pixelation effects, but beginning with the 16 pix/f level the drop in accuracy was more substantial. Visemes /tha/ and /la/ were exceptions in that a conspicuous decrease in accuracy had begun already with 64 pix/f quantization. Campbell and Massaro (1997) noted that performance in visual speech reading with pixelated dynamic images is similar to face identification in terms of robustness of perception despite stimulus degradation.

Auditory perception of speech is substantially influenced by visual perception of the visemes accompanying the uttered sound of the syllables. Robustness of this intermodal influence is persuasively demonstrated by the McGurk and MacDonald effect (McGurk & MacDonald, 1976). When certain recorded audible and videotaped syllables are artificially constructed so that the auditory stimulus is accompanied by a nonmatching viseme, illusory perception is produced. For example, when an audible /ba-ba/ is accompanied and synchronized by the visual clip of a face uttering "ga-ga", illusory "da-da" is heard indicating that a fusion effect is obtained. MacDonald et al. (2000) examined whether the McGurk effect tolerates pixelation of dynamic facial images uttering the stimuli syllables. Videotaped clips of a model uttering syllables typically used in the McGurk effect studies were prepared so that congruent and incongruent audiovisual stimuli could be used in the main experiment. Pixelation level as the main independent variable was changed from no pixelation control condition to 30 pix/f (12 pix/mouth), 19 pix/f (7.7 pix/mouth), and 11 pix/f (4.4 pix/mouth). Surprisingly high tolerance of the illusory effect to substantial pixelation was found. For example, with spatial quantization level at 19 pix/f the mouth area of a dynamic facial image consists of only 7.7 pixels and at the coarsest pixelation level the mouth is depicted by as few as about 4.4 pixels. Despite that with this coarse pixelation mouth as an element of a face could not be recognized (when presented in isolation), the illusion persisted with most of the incongruent audiovisual pairings of phonemes and visemes that typically lead to the illusion in the usual conditions without pixelation. Consequently, the effect of visual cues on auditory fonemic perception should be based on coarse-scale spatial information and possibly on certain cues extracted from motion of the mouth region of the facial image. (The most robust illusory effect in the condition of the coarsest pixelation levels still allowing the illusion was obtained with syllables visual /ga/—auditory /ba/ producing illusory "da", and also visual-auditory pairings /ka/−/pa/, /da/−/ba/, /ta/−/pa/, /ma/−/na/.)

Munhall, Kroos, Jozan, and Vatikiotis-Bateson (2004) studied audiovisual speech perception capitalizing on band-pass filtered SF images of dynamic faces. Center frequencies of the filter ranged from

2.7 to 44.1 cycles per face image (rough equivalents of 5.4–88 pix/image, respectively). The facilitating effect of visual information on auditory intelligibility for speech, which was presented in noise, was found with a widely varying spectral content of the dynamic images. Compared to the audio-only condition, audiovisual presentation increased word recognition rate by 10–20%. The mid-range SF content (equivalent to 22 pix/f) proved to be most effective. The effects were invariant to viewing distance indicating that not absolute level of SF, but cycles of contrast per facial image mattered. Similarly to the findings by MacDonald et al. (2000), Munhall et al. (2004) concluded that HSF information is not necessary for audiovisual speech perception. They also point out that a limited range of SF spectrum is sufficient for the audiovisual methods of target image masking effects.

CHAPTER 9

Pixelated Images Used as Masking Stimuli

In some sense, the pixelation transform can be considered as one of the methods of masking. The mosaic of square-shaped areas with uniform brightness and sharp edges and depicting small orthogonal corners of these areas essentially acts as noise interfering with original image content. Furthermore, beginning with a certain level of increasing coarseness (ie, increasing size of the pixels) this transform produces a kind of new object—a block mosaic—that competes with original image content preserved in the configuration of spatial contrast gradients carried by the low spatial-frequency (LSF) contents. This makes another aspect of masking—the between-objects masking. However, in case of traditional approaches to pixelated image perception masking aspects and masked aspects are always temporally entangled, simultaneously present in the stimulus image. As the temporal aspect of target–mask interaction has always been the key factor in predicting variability in target perception, it would be useful to develop pixelation-based research so as to disentangle spatial and temporal variables. Therefore, pixelated masking images derived from target images can be used as masks for the targets, while temporal separation between them is varied. In addition to this, there are other advantages of pixelation as a means of masking compared to the more often used traditional degradation/masking methods. Only a handful of experiments have been done with these considerations in mind. One such study was carried out by Bachmann et al. (2004).

Bachmann et al. (2004) explained the theoretical and methodological underpinnings of their approach as follows (I will rely on their text from pages 12 to 15 quite literally). Recognition and identification always involve generalization over the variable concreteness of the unique patterns and images of objects and scenes. Generalizations leading to an invariant perceptual specimen or category are the essence of recognition. But the problem is that we do not know how to quantify similarity or difference between images in a perceptually meaningful

Perception of Pixelated Images. DOI: http://dx.doi.org/10.1016/B978-0-12-809311-5.00009-X

manner (Sinha, 2002a, 2002b). It is obvious that a main strategy for studying why the mind is so versatile in pattern recognition should involve the use of well-quantifiable degradations and transformations of stimulus images in order to find the limits and the sources of the constraints of invariance-extraction skills. In order to study the relative roles of different stages in information processing, the experimental manipulations of the quality of information at these stages should be guaranteed. Wholistic inversion, spatial-frequency (SF) filtering and masking by noise have been the standard methods used for this purpose, but each of these methods has its own methodological limitations. Wholistic inversion is a manipulation that is typically used for the control of configuration processing, but in this case one does not manipulate exactly the configuration as such—the relative locations of the features within the image structure. Simply, the conditions are provided that make it more difficult for perception to use this information (Rakover, 2002). Low-pass SF filtering at a progressively lower cut-off value of SFs eliminates image detail so that information left in the image will be progressively less detailed and more coarse. However, subjects are very efficient at recognizing the identities even if the images are low-pass filtered in the Fourier domain or by some Gaussian coarsening operations (Sinha, 2002a, 2002b). Despite SF filtering, observers have a representation of the wholistic configuration sufficient for good recognition (eg, Goffaux, Gauthier, et al., 2003). When invariant visual noise masks are used to control the quality of information available from the target image, local features can be obscured and the signal-to-noise ratio of features, SF spectra, and spatially localized tokens for wholistic configuration decreased. Yet the basic configurational characteristics are not directly manipulated. This is a shortcoming also because configurational characteristics pertaining to interfeature distances differently influence performance in higher-level face recognition and lower-level facial image perception tasks (Schwaninger, Ryf, & Hofer, 2003). However, if degradation by blocking-pixelation as applied directly to target stimulus images is used, several sources of image recognition are eliminated or impoverished at once and it will be difficult to attribute the causes of the effects specifically to one or another stage of perceptual image processing. This is the prime methodological weakness inherent in straightforward spatial quantization of the test or target images (see also Morrison & Schyns, 2001; Schyns & Oliva, 1999). It is thus important to keep target images unchanged between different experimental conditions in order to have the invariant and full target-specific perceptual data as the input to the processing system. Visual masking is a method that allows us both to keep target images intact and to introduce perturbations into different microgenetic stages of information processing (Bachmann, 1994, 2000; Breitmeyer, 1984).

When two spatially overlapping images are presented in rapid succession, the first image (S1) can have an interfering as well as a facilitatory/supportive effect on the processing of the following image (S2). If the images are identical, the supportive effect is maximal because all important aspects of the second image processing are facilitated by the preliminary processing of the previous image: (1) spatiotemporal integration of local sensory signals is amplified and the signal-to-noise ratio of the local features as well as overall configuration of the features is emphasized; (2) the search for the perceptual category of the object is initiated ahead of time by the first image and as the two images are identical, this operation becomes redundant or considerably simplified (facilitated) for the second image; (3) the first image acts as an object that captures attention and processing of the second image signals benefits in terms of an increase in the perceptual saliency and speed with which these signals are processed. In the opposite case, if images are cardinally different, the first image: (1) obscures the features of the second image and reduces the contrasts of the spatial layout of the image at the integrative stage of processing (decreases dramatically the signal-to-noise ratio); (2) sets the search for the perceptual category off in a wrong direction, thus necessitating the initiation of a new categorical search for the succeeding image; (3) puts a higher demand on the limited-capacity attentional system. Target images per se remain nondegraded; however, the contaminating factor will be introduced with the help of forward-masking images. The function of forward masks will be primarily twofold: First of all, to allow the introduction of theoretically meaningful, parametrically varied manipulations into the integrative stage of the processing of target images (that follow the mask); secondly, to provide the input for microgenesis to start with, in order to see in what conditions the processing of the initial input from the masker stimulus may constitute the stage of processing that can be considered as a preprocessing of the following target stimulus. The view that the first stimulus starts a common processing so that when the second stimulus arrives, some preprocessing for it has already been carried out as based on the first stimulus has been accepted in some studies on masking in which faces were used as stimuli (Bachmann, 1989; Calis et al., 1984; Costen et al., 1994). When the following stimulus is presented, processing does not begin anew, but continues from the "half-way" point, where the processing that is dealing with some aspects of the first stimulus is taken over by the processing that deals with the second stimulus. Interstimulus compatibility that permits interactive microgenesis is provided by overlapping location, compatible size, and some robust and shared configural aspects of the stimuli. If it is important to introduce masking noise into a face recognition system at the early integrative stage where spatial filtering operations are performed before the categorical and semantic levels of processing (Morrison & Schyns, 2001), we should use forward masking, instead of the more often used backward masking. This is because forward masking has its effect primarily at the early integrative stage of visual processing (Bachmann, 1994; Breitmeyer, 1984; Breitmeyer & Öğmen, 2000; Turvey, 1973). This is especially important in the use of faces as stimuli because in face recognition operations, individuation of a face strongly depends on original spatial filtering compared to what happens with the use of other objects (Biederman & Kalocsai, 1997).

Bachmann et al. (2004) aimed at comparing the predictions of the different theoretical accounts of face recognition on the level of masking as dependent on the type of the forward mask. The local feature processing account (FP) puts responsibility for successful face recognition on local features specific to faces. The generalized SF processing account assumes that recognition is based on extracting the SF-content from the image and as this content differs between individual images, these images can be specified. The theory of microgenetic processing of the wholistic configuration (MP) says that image processing in perception is a stage-wise succession of operations, beginning with coarse-scale configural representation and gradually adding detail to this proto-object stage.

In the experiment by Bachmann et al. (2004), all masking stimuli were spatially quantized over a systematically more coarse scale of pixelation, which allowed relatively different effects on the different levels of the processing of the target-image attributes. By transforming masker image configuration the impact of the masker on the target's configuration processing is changed. Presenting local masking noise carried by the mosaic of the blocks of a quantized image helps interfere with local feature processing. Importantly, in order to take control over the covarying changes in SF content, the amount of masking noise, the distortion of the local features, and the distortion of configural information (all these effects accompany changes in the scale of quantization), Bachmann et al. (2004) used different types of prequantization images to prepare forward masks. (1) In one type of premask, its source image replicated the target both at the local features and configurational levels. In this case a replica of the target is used as the source image, but pixelated at different levels. (2) The other type of premask replicated the target only at a general (robust) configurational level common to the same category of objects (eg, faces, but was different in terms of local features and subtle spatial characteristics of configuration). In this case a face different from the target was used and pixelated at variable levels of spatial scale. (3) The third type of premask was used in order to have a control over the SF content apart from the configural information. For this purpose visual broadband Gaussian noise with power spectra similar to the spectra that are typical for faces was used. This type of mask contaminates spectrally tuned SF analysis mechanisms of the target. However, it still permits some information about the spatially nested configuration of the target because of larger differences in the spatial phase of contrast between the noise mask and the target image compared with the difference between the spatial phase of contrast of different face mask and face target. Importantly, as spatial quantization eliminates information about the form of local features rather than about the mutual distances and directions between the approximate locations of them, it should be considered as a transformation that impairs the fine-scale relational aspect of configurational processing (eg, see Leder & Bruce, 2000; Schwaninger et al., 2003) rather than the coarse-scale wholistic aspect of configurational processing.

Several predictions were put forward (Bachmann et al., 2004) with regard to the expected experimental results, depending on the stimulation conditions. It was assumed that when the quantized mask before its pixelation was the replica of the target and when pixelation is fine-scale (ie, very small blocks making up the quantized image of the mask), target information at the local features level, as well as the wholistic configuration level, is supported by the mask image and the integrative, entry level of processing will emphasize target information. Because the process of category search for the stimuli is redundant or doubled, the mutual support between mask and target should be amplified further. Because a scale of pixelation higher than 16—18 pix/f (horizontally) allows the identity of local features to be communicated (despite some local fine-grained noise), the feature-based identification process should not be compromised. With intermediate-scale pixelation (10—16 pix/f rendering blocks of intermediate size) only wholistic configural information must be supported by the mask structure, but local features will be masked by local block mosaic including horizontal and vertical edges, squares of uniform brightness, and orthogonal corners of the mosaic blocks (see Fig. 9.1 S1 upper left for an example). The local detailed

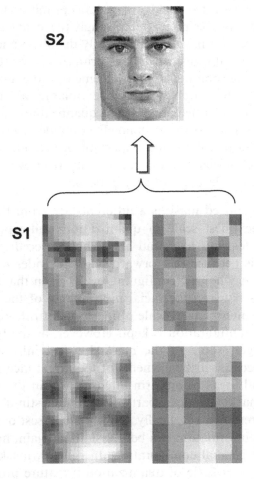

Figure 9.1 Examples of forward masks (S1) and a target (S2) from Bachmann et al. (2004). S1 could be the same face as the target, a different face, or random Gaussian noise; pixelated at different levels of coarseness.

structure will be masked and distorted while wholistic configurational information that is invariant to local detail (a "face primitive") will be largely saved, its minor distortion notwithstanding. (If we would extract only the eye or mouth area from Fig. 9.1 S1 upper left, these individual features could not be recognized in isolation.) Fine-scale and intermediate-scale pixelation should enable the correct recognition due to a mutually supportive integration of mask and target attributes and to confusion of the correct responses to the mask with the correct responses to the target. With the coarsest level of pixelation (eg, 5–10 pix/f; see Fig. 9.1 upper right for an example), also the wholistic configural structure of a particular face will be distorted and facial identity lost. On the other hand, the general primitive for the category of faces as such can be preserved through the coarsest quantization, although local features have become totally dissolved and configuration of a particular face also distorted. This primitive supports processing of whatever shares the robust spatial arrangement of the image characteristics resembling a "face archetype," while violating all featural identity information. If an intermediate scale of quantization used in the same face masking has a strong detrimental effect on identification, it must be concluded that local features are important in face identification. If an intermediate level of quantization has a weak effect, we should conclude that the local feature level is not very important.

When the quantized mask is a different face from the target in the condition of the finest scale of quantization well-expressed masking both at a local features level and at a wholistic configuration level is expected. This is because the forward mask provides target distortion at local as well as wholistic configural levels, given that there are sufficient differences between the facial morphology of the target and the mask. With the intermediate scale of mask quantization, the distortion of features and configuration is kept. However, as the blocks of quantization get larger, some of the masked target information may be saved. This is because if the elements of the target face image are integrated with local areas of uniform intensity within the relatively large blocks of the masking image, part of the target stimulus' local structure will be represented veridically. With the coarsest quantized masks, fine local target information can be transmitted again, however disintegrated from the general configuration. In general, masking should get stronger with a finer scale of quantization if feature processing and/or configuration processing is the principal basis of identification.

When the quantized mask is derived from random broadband Gaussian noise with the power of the components mimicking the power spectrum of a typical face image, the following is expected. With the finest scale of quantization, local features, global features, and SFs that are typical for faces are quite well masked; however, there will be relatively weaker masking competition for a wholistic configuration that is typical for a face compared with the respective effect with different-face masks (see Fig. 9.1 S1 down left for an example). This is because random noise does not contain well-articulated facial configuration. With an increase in coarseness of quantization applied to the mask image (up to the coarsest levels), the masking of local features should gradually diminish because: (1) areas of uniform intensity become larger, (2) there is still no configurational structure that would compete well with spatial structures that are typical for the generalized configuration of face elements. If this type of mask provides masking that is strongly dependent on scale of quantization, either local feature processing or SF analysis have to be crucial mechanisms of face identification. If, however, masking is independent of coarseness of quantization, wholistic configural processing mechanisms are of prime importance for face identification in our experimental conditions. [In the study by Bachmann et al., a simple identification task was used and therefore the issue of task specificity and diagnosticity of different aspects and scales of facial information was avoided. Relative roles of features and spatial scales as the basis of diagnosticity may change depending on tasks. For respective discussion see Schyns, Goldstone, and Thibaut (1998), Schyns and Oliva (1999), Morrison and Schyns (2001), Oliva and Schyns (1997), Gosselin and Schyns (2002), De Gelder and Rouw (2001), Rakover (2002), and Goffaux, Jemel, Jacques, Rossion, and Schyns (2003)].

In the experiment, premasks were presented for 82 ms spatially precisely overlapping with targets which were presented for 23 ms with stimulus-onset asynchronies (SOAs) ranging between 82, 105, 128, 152, and 175 ms. [inter-stimulus intervals (ISIs) were, respectively, 0, 23, 46, 70, and 93 ms.] Eight-alternative forced identification task was used. Accuracy significantly depended on pixelation level, type of mask, and SOA.

Face identification improved with increase in SOA, moving from fine-scale to coarse-scale quantization, and was more susceptible to masking with quantized images derived from faces than to masking with pixelated noise. A highly significant interaction between SOA and type of mask indicated that while the same-face mask had only a marginal masking effect at all SOAs, different-face mask and noise mask had progressively stronger masking with decreases in SOA, indicating that processes at the early integrative stage of facial image processing were also involved. There was a strong interaction between type of mask and quantization level. The same-face masks did not cause any substantial masking with any level of quantization, but masking was stronger with a decrease in the level of coarseness of pixelation both with different-face mask and noise mask. With fine-scale quantization, different-face mask impaired identification more than noise mask; with coarse-scale quantization, the opposite was true. Fig. 9.2 shows these effects. This pattern of results contradicts local feature processing and generalized SF processing theories of face recognition, but supports a microgenetic configuration processing account. The cross-over of identification functions in the face mask condition and noise mask condition shows relative facilitation of target identification by face mask with pixelation levels less than 11 blocks per face.

Interestingly, 11 pix/f level is similar to the value of pixelation that specified a step that was found to be critical for an abrupt decrease in face identification in earlier studies (Bachmann, 1991; Bachmann & Kahusk, 1997; Bhatia et al., 1995). It is likely that pixelation at 16 pix/f represents a border case between depiction enabling face individuation and depiction carrying only general face primitive shared between faces

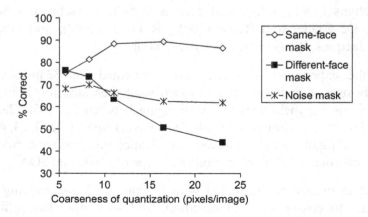

Figure 9.2 Accuracy of face identification (% correct) as a function of level of pixelation and types of premasks (adapted from Bachmann et al., 2004).

with different identities. The 11 pix/image may be a borderline condition between general noise and configural noise effects—a further decrease in pixel size adds configural noise to general broadband noise. With most-coarse pixelation conditions the same-face mask and the different-face mask both allowed equally efficient perception at a level of performance higher than that with the noise mask (Fig. 9.2). This result is consistent with the microgenetic configuration processing theory: despite a difference at the identity level of processing, different faces may share the processing routines for the general visual category of face-like noisy configurations at the microgenetic early stage of proto-object encoding.

Same-face masks did not have strong masking effects at any of the pixelation levels, including the coarse level where identity is lost, but proto-object representation sustained. Thus, feature-processing theory (FP) was not supported, but this result was consistent with SF and microgenetic configuration-processing (MG) theories. Different-face masks exerted progressively stronger masking effect with a progressively finer scale of pixelation and all three theories are consistent with this result. Quantized Gaussian noise masks had roughly the same effect across all quantization levels, but masking increased with a decrease in SOA. Feature-processing theory and the SF theory cannot explain this result, whereas the microgenetic configuration processing theory fits well.

In general, pixelated mask images originating from facial images that are different from targets were the strongest type of mask in this paradigm (see also Fig. 9.2). This was especially pronounced with coarse scales of pixelation and shorter SOAs (see Fig. 9.3).

The various experimental data presented by Bachmann et al. (2004) supported occasionally any of the three theories of face identification, but the only theory not contradicted by any of the particular results of that study was the microgenetic configuration-processing theory. (See Table 9.1 which is presented in order to summarize this conclusion.) According to this theory, "... at the initial stages of image processing, a coarse universal representation is formed for any form that slightly resembles a face (without information for identity involved as yet). Subsequently, this representation is updated and/or replaced by a configurational representation that is locally refined, with individuated values of the metrics of particular faces specified as well as local features added to coarse wholistic configuration. (This explanation is

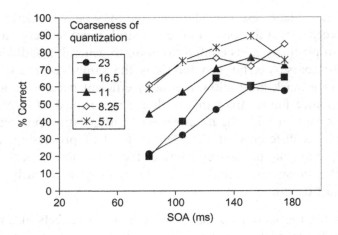

Figure 9.3 Accuracy of face identification (% correct) as a function of SOA between premask and target and level of pixelation in the condition of masks derived from faces different from the target (adapted from Bachmann et al., 2004).

Table 9.1 The Consistency of Different Theoretical Accounts in Explaining Empirical Findings (Table 2 From Bachmann et al., 2004)

	Expected and Actual Level of Masking					
	FP		SF		MG	
Type of Mask	Theoretical	Factual	Theoretical	Factual	Theoretical	Factual
Same-face						
Coarsely quantized	No	No	No	No	No	No
Intermediately quantized	*Yes*	*No*	No	No	No	No
Finely quantized	No	No	No	No	No	No
Different-face						
Coarsely quantized	No	No	No	No	No	No
Intermediately quantized	Yes	Yes	Yes	Yes	Yes/no	Yes
Finely quantized	Yes	Yes	Yes	Yes	Yes	Yes
Noise-mask						
Coarsely quantized	No	No	No	No	No	No
Intermediately quantized	*Yes*	*No*	*Yes*	*No*	No	No
Finely quantized	*Yes*	*No*	*Yes*	*No*	No	No

Note: Instances of inconsistency between the theoretical predictions and empirical factual findings are indicated in italic.

similar to several microgenetic accounts of processing that accept representational progression from coarse wholistic [general categorical] to fine detailed [individuated] levels within the first 20−200 ms of image processing: Schyns and Oliva, 1999; Oliva & Schyns, 1997; Parker,

Lishman & Hughes, 1992, 1997; Sanocki, 1993; Hughes et al., 1996; Navon, 1977; Liu, Harris & Kanwisher, 2002.) Our results support the notion of quantization as a type of degradation that primarily interferes with configuration processing. Moreover, among the three operations —feature processing, spatial-frequency analysis, and configuration microgenesis—the configural one appears decisive in processing quantized images and in terms of the principal effects brought about by quantization." (Bachmann et al., 2004, p. 18). It was proposed that in the forward-masking paradigm the masks act both as setting the preliminary perceptual context for the initial perceptual operations and as interfering images, the spatial content of which is used to contaminate the initial representation at the integrative entry level of spatiotemporal information processing. From the methodological point of view it is worth noticing that at each level of pixelation, the different types of mask images had mutually compatible SF content and coarseness of the elements that represented the generalized spatial structure. What was essential, was that these types differed in terms of configurational similarity to target images.

For example, the intermediate level of mask pixelation used in the same-face masking condition did not have a strong effect and did not differ in terms of effect from other levels of quantization. Consequently, the listing of individuated local features as a possible recognition mechanism is not the most important subprocess in face recognition. Also, some sort of counterbalancing between the effects of different sources of identity may be present: at the local features level the intermediate quantized same-face mask provides strong interfering features, but also a supportive wholistic structure for the target image. In the different-face masking condition, the coarser the pixelation, the weaker the masking. Feature theories and configurational theories of face recognition are both consistent with this. However, when instead of a different-face mask a noise mask was used, there was no substantial difference in masking dependent on coarseness of pixelation. This supports the wholistic configuration processing mechanisms as the crucial ones in face identification. Because in the forward version of masking primarily the early integrative stage of processing is tapped (Breitmeyer & Ögmen, 2006), the configural processing should be immediately dependent on the spatial layout of local feature contrasts.

It is only logical that in addition to forward masking by pixelated masks backward masking should also be studied. This was done by

Bachmann and colleagues in a related paper in a different journal (Bachmann et al., 2005b). The stimuli were the same as shown in Fig. 9.1, but the order of presentation was reversed—the unpixelated target was exposed first (for 23 ms), followed by the pixelated mask (83 ms). Values of SOAs varied between 23, 46, 70, 93, and 117 ms. As the main objective, the authors extended the strategy of varying the contents of masks to backward pattern masking where targets and masks overlap in space, in order to compare different masking theories.

Bachmann et al. (2005b) advanced the following hypotheses. (1) With a Gaussian noise mask as a "nonobject" there will be less attentional competition between target and mask (as two identifiable objects) compared to different-face masks condition. Thus it was hypothesized that there would be considerable release from masking at long SOAs when the noise-mask is used compared to masking with different faces. (2) With different-face masks and Gaussian noise masks fine scale of pixelation will lead to stronger masking compared to coarse-scale pixelation, because fine-scale masks provide a more metacontrast type of interference at the local image areas. (3) Because at the same level of spatial quantization the generalized SF content of different masks is kept the same (while configuration was varied), differences in the degree of backward masking between different masks at intermediate-to-longer SOAs (implicating interchannel interaction) could not be expected if transient-on-sustained theory was the sole way to explain backward masking in these conditions. (4) Because masks which have a facial configuration provide stronger capacity for attentional capture compared with noise masks, stronger masking is expected by different faces than by noise if attentional object substitution theory was a valid explanation. (5) As masking at the shortest SOAs depends mainly on intrachannel inhibition when S1 and S2 are integrated, noise masks as well as different-face masks must produce strong masking and must be equally influential while same-face masks are expected to produce weak masking at the shortest SOAs.

The main results were as follows (Bachmann et al., 2005b): Configural characteristics, rather than the spectral content of the mask, predicted the extent of masking at relatively long SOAs. This is difficult to explain by the theory of transient-on-sustained inhibition (Breitmeyer, 1984) as the principal mechanism of masking. The backward-masking theories that assume strong sensitivity of masking mechanisms to the SF content of the stimuli did not receive conclusive support. Local contour interaction also hardly explains masking in these conditions. The scale of quantization had no effect on the masking capacity of noise masks but showed a strong effect on the masking capacity of different-face masks. The decrease of configural masking with an increase in the coarseness of the quantization of the mask refers to ambiguities inherent in the re-entrance-based attentional substitution theory of masking. Different masking theories each were incapable of solving the problems of masking separately. It was concluded that the theories should be combined in order to create a complex, yet comprehensible, mode of interaction for the different mechanisms involved in visual backward masking.

Figs. 9.4–9.6 illustrate the main results. The robust difference between different-face mask effects (strong masking with intermediate-to-fine pixelation) and noise-mask effects (equal masking at all pixelation levels) refers to configuration-processing mechanisms mainly involved in masking; local feature-interference effects and general SF content-based masking effects seem not to play an important role. Backward masking at long SOAs depends on objectness and configuration-based attributes and not on SF-noise or local lateral interference.

Based on the experimental results (Bachmann et al., 2005b), the authors discussed several theoretical implications of their study, but stressed that the whole set of empirical findings in toto, especially

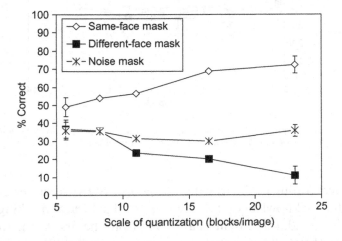

Figure 9.4 Accuracy of target-face identification as a function of type of mask and scale level of mask pixelation in the backward masking conditions (adapted from Bachmann et al., 2005b). With coarse-scale pixelated masks different types of mask have comparable effects; with progressively finer levels of mask pixelation masks originating from the same faces as targets, allowed systematically better target perception, while masks originating from different faces lead to an increase in masking.

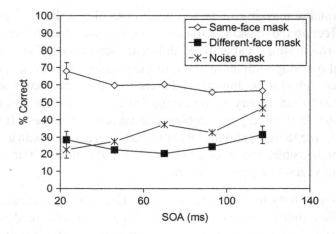

Figure 9.5 Accuracy of target-face identification as a function of type of mask and SOA in the backward masking conditions (adapted from Bachmann et al., 2005b). The masking effect of the Gaussian noise mask becomes weaker with SOA; the strong masking effect of different-face masks sustains over all SOA values.

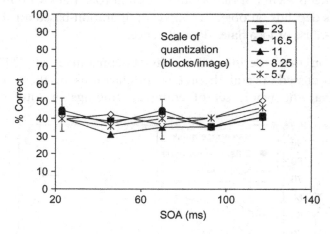

Figure 9.6 Accuracy of target-face identification as a function of level of pixelation and SOA in the backward masking conditions (adapted from Bachmann et al., 2005b). SF content in general does not have an effect on masking.

when concentrating on intermediate-to-longer SOAs, requires an explanation that acknowledges mechanisms of backward masking which include the mandatory attentional switch from the first to the following integrated object. They noted that a more complex theory involving attention is necessary even without typical substitution-masking conditions (such as distractors and/or common-onset and asynchronous offset displays).

Strong visual masking occurs also when target and mask are presented to different eyes, called dichoptic masking (Bachmann, 1994; Turvey, 1973; Werner, 1940). This experimental variety shows that central interactions between the competing stimuli are where masking originates. This type of masking is a "family-resemblance" of another perceptual suppression phenomenon called binocular rivalry. When sufficiently different images are presented to different eyes, even for a longer duration, contents of explicit perception begin to alternate between the right and left eye stimuli. (For the exclusive dominance and suppression so that all input from one eye is dominating over the whole input presented to the other eye without blending and alternate interocular "patching," the images have to be relatively small.) Dichoptic effects of pixelated images on target-image perception were studied by Bachmann and Leigh-Pemberton (2002). In the following a brief overview of that study is presented.

Unpixelized images of faces were presented to one eye and the pixelized versions of the same faces to the other eye. This paradigm has several advantages because low-level luminance and general object category of the rivalrous images are the same and the number of contributing variables thus better controlled. The level of pixelation was varied between eight values from 5 up to 78 pix/f. The time a pixelated version was dominant over its unpixelized counterpart increased with increasing coarseness (ie, increasing block size) (see Fig. 9.7). The authors pointed out that whereas all rivalrous stimuli (1) were derived from the original images that belonged to invariant perceptual object category and had invariant exemplar *identity*, (2) had equal and invariant overall *luminance*, and (3) were characterized by the invariant set of (vertical and horizontal) *contour orientations* of the edges of elements of the quantized images then the main determinants of dominance in rivalry should have been related to the differences in SF content and/or wholistic pattern configuration of the rivalrous stimuli. Whereas the meaningfulness and ease of interpretation (face-like quality) of the image decreases with coarseness of pixelation then the concomitant increase in rivalry dominance must not depend on high-level categorical or identity processing but on some intermediate-level processes where the physical-configurational (Gestalt-) description of the image is sought for.

Several comments are worth mention in the results of Bachmann and Leigh-Pemberton's (2002) exploratory study. From earlier observations

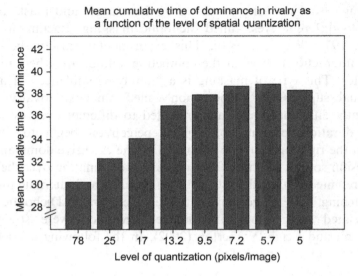

Figure 9.7 Mean cumulative time (s) of exclusive perception of rivalrous pixelated images as a function of the level of pixelation. The coarser the level of pixelation, the longer the time the stimulus is dominant in rivalry. Notice the leveling off of the increase in dominance after the level of quantization has reached 13.2 pixels per face image horizontally. (Graph adapted from Bachmann and Leigh-Pemberton, 2002.)

it was known that meaningfulness, category-likeness, and interpretability tend to increase perceptual dominance in rivalry (Uttal, 1981; Walker, 1978). This allows prediction that the coarser the level of pixelation (and, therefore, category-likeness and interpretability) the less the perceptual dominance. The results were exactly opposite. It is known that the amount of contour and/or texture positively correlates with rivalry dominance (Breese, 1899; Levelt, 1968). This also predicts that original images and fine-pixelated images should dominate. Again, the results showed just the opposite. The increase in dominance of the coarse-pixelated images could be best interpreted as a result of the processing routine biasing perception towards precategorical figural Gestalt formation regardless of the identity and/or meaningfulness of the image. The coarser the level of pixelation, the more the configuration of the pixelated image differs from the wholistic configuration of the unpixelated image of the same object and for some reason this tends to lead to increased dominance. Also, the coarser the pixelation level, the smaller the number of well-articulated elements constituting the image and the higher the relative share of LSF content in the image compared to the high SF (HSF) content. For some reason these factors may be beneficial for perceptual dominance.

CHAPTER 10

Effects of Pixelation on the Electrophysiological Signatures of Face Perception

Most of the research on how people perceive pixelated images is behavioral studies using psychophysical methods. Although the subjective and objective behavioral side of the problem are both important, we may want to know also what the neural correlates and psychophysiological signatures of perception are when stimuli consist of spatially quantized images of objects, scenes or people. On the one hand, studying neural correlates of pixelated image perception theoretical models of perception can be constrained by objective neurobiological data. On the other hand, there can be circumstances when perceivers are either reluctant to describe their experiences veridically or they may be in a state—either normal or pathological—where verbal or other typical means of communication cannot be used. In this context, knowing more about neural correlates of the pixel-image perception may be highly important theoretically as well as from the applied perspective.

Up to now only a very few attempts have been taken to follow this direction of research. Let me introduce these. (All of this has been carried out by EEG in order to register and measure event-related potentials—ERPs—online with pixelated image perception. It has been well known for some time that certain specific ERP signatures are sensitive to face-image stimuli, including the low-pass spatial frequency (SF)-filtered images—for example, Allison et al., 1994; Bentin, Allison, Puce, Perez, & McCarthy, 1996; Eimer, 2011; Flevaris, Robertson, & Bentin, 2008; Goffaux, Gauthier, et al., 2003; Goffaux, Jemel, et al., 2003; Halit, de Haan, Schyns, & Johnson, 2006; Holmes, Winston, & Eimer, 2005. However, using pixelated images for face-specific EEG-based research has been rare.)

In one such study Hanso et al. (2010) engaged observers in playing a "game": they were told that researchers try to use EEG to retrospectively predict which faces were familiar to them and which were not. Thus they had to look at each presented face, but respond always by

Perception of Pixelated Images. DOI: http://dx.doi.org/10.1016/B978-0-12-809311-5.00010-6

saying "unfamiliar" in order to deceive the researchers. Pixelated face images (3.8 × 5.7 degrees, three levels of pixelation—6, 11, 21 pixels/face) were shown for 480 ms each and ERPs were recorded from occipital, temporal, and temporal-parietal (TP) electrodes (in the latter case the region of interest included TP7, TP8, P3, and P4). Among the 12 alternative faces, 2 were familiar to the observer. A double-blind experimental protocol was used. Among several ERP components known to be sensitive to different aspects of facial image perception, N170 and P300 were analyzed. Results showed that (1) pixelated face images, when perceived, produce N170; (2) the amplitude of N170 decreased with an increase in the coarseness level of pixelation; (3) there was an interaction between face familiarity and level of pixelation in determining the P300 amplitude—with unfamiliar faces P300 was equally expressed across different levels of pixelation but with familiar faces P300 amplitude became progressively larger with an increase in the number of pixels per face (ie, with fineness of detail). From Fig. 10.1 one can see the basic results as expressed by ERPs recorded from the TP electrodes. We notice that even though a picture of a face is pixelated, it leads to a face-typical N170 with a rather high amplitude unless its pixelation level is too coarse (eg, 6 pixels/face which is an approximate equivalent of 3 cycles/image). This is with both, unfamiliar and familiar faces. We can more or less safely conclude that from 11 pixels/face up, spatially quantized images of faces produce N170. We also notice that the P300 amplitude systematically increases with the number

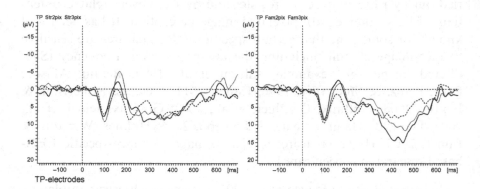

Figure 10.1 Grand average ERPs registered from temporal-parietal pooled electrodes (negativity up). Distinct P100, N170, and P300 can be seen. Recordings on the left—ERPs to unfamiliar faces; right—ERPs to familiar faces. For 21 pixels/face images condition ERPs drawn in black; for 11 pixels/face ERPs in red (dotted line in print versions); for 6 pixels/face ERPs in green (dashed line in print versions). Source: Adapted from Hanso et al. (2010).

of pixels per image for familiar faces, but not for unfamiliar faces. However, due to the special "catching a deceiver" type of experimental design we cannot say whether the cognitive factors or emotional factors related to cognitive ones play a leading role in the P300 effect here.

The results by Hanso et al. (2010) suggest that in the forensic context, where eyewitnesses sometimes may be asked to recognize suspects from pixelated images (and especially when familiarity of the person could be a factor), their responses could be trusted, provided that images are not pixelated below a certain benchmark level of coarseness (eg, 11 pixels/face). Another way to capitalize on these results would be to try to get hints about whether a person actually knows somebody personally, but has decided to conceal this knowledge.

Recently, in collaboration with Kevin Krõm, Geidi Sile, Renate Rutiku, Carolina Murd, and some other students, we have obtained some additional data on the effects of pixelation on ERPs. When facial images sized and pixelated similarly to what was done by Hanso et al. (2010) were used in a different task instructing observers to report whether the image was an animate or inanimate object (pixelated faces, living creatures, and inanimate objects were presented for 450 ms), ERP signatures were sensitive to faces again. (Fig. 10.2 gives an example of one of the inanimate objects used in the study.) First, this experimental protocol allows to free observers' attention from selectively expecting the objects belonging to the category of faces. Second, this task is free from perceptual sets or task demands in which subjects are aware that person identification is the key task, which enhances the study of EEG signatures of automatic person perception based on pixelated images.

Figure 10.2 Examples of an inanimate object's images pixelated at different levels of coarseness.

Results showed that N170 amplitudes for unfamiliar as well as familiar faces were well pronounced in the 11 and 21 pixels/image conditions, but had a small amplitude when 6 pixels/image facial stimuli were displayed. (TP electrodes were combined with O1 and O2 electrodes in this case.) Inanimate objects, even though they also produced a N170-like ERP component, did not show differential potential level as a function of pixelation coarseness. In this sense, the N170 component can be trusted as a face-perception signature if it shows sensitivity to the level of pixelation. I believe that, compared to the traditional SF-filtered images of faces in ERP research on face perception, pixelated images have certain advantages. Standard low-pass SF filtering simply removes higher frequencies from the image, but face resemblance does not suffer much from this. However, with a pixelated image face resemblance degrades considerably by spatially quantizing the image over progressively coarser scales. This means that ERPs can better differentiate face-category versus nonface stimulus category processing as the prevailing operation the brain does in response to a certain stimulus image.

In another experiment, images depicting one of the two different categories (face or house) were shown to the observers for just 10 ms. Unknown to the participants, the first category had two subcategories (familiar face or unfamiliar face). The stimulus material was pixelated at four different levels (1–14, 2–11, 3–8, 4–4 pixels/image) and the observers had to answer which of the two categories (face or house) they had seen on the display. Behavioral results are drawn in Fig. 10.3. Familiar faces were detected better than unfamiliar faces on every pixelation level. On the first three levels of pixelation (representing finer scales) familiar faces were detected better than houses, but no differences appeared between the detection of unfamiliar faces and houses. However, the coarsest level of pixelation caused face categorization to drop dramatically, while house categorization decreased only moderately—it is likely that with large image blocks face-likeness is destroyed more than house-likeness. Behavioral effect for the familiar faces was echoed also in the EEG data. Differences between these categories appeared about 200 ms poststimuli, which is similar to other studies examining the speed with which familiar faces are discriminated (Caharel, Ramon, & Rossion, 2014). As can be seen from Fig. 10.4, 14 and 11 pixels/image pictures allow discrimination of familiar and unfamiliar faces by the brain (N170 amplitudes differ and are higher for familiar faces), but more coarse pixelation levels obscure, if not reverse, the familiarity effect, although at

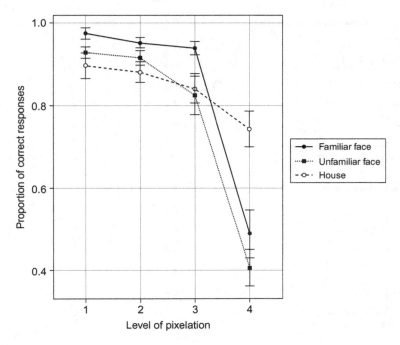

Figure 10.3 Performance in the object categorization task (proportion of correct responses) as a function of object category (familiar faces, unfamiliar faces, houses) and level of pixelation.

the 8 pixels/image scale of coarseness seeing faces still leads to higher amplitude of the potential compared with houses.

This study again shows that the critical range of the level of pixelation coarseness beyond which category-specific ERP components lose sensitivity of discrimination is specified by the 11−14 pixels/image level. Interestingly, this level quite well corresponds to the values of pixelation shown to be critical in the behavioral face identification tasks (eg, Bachmann, 1991; Costen et al., 1994). It is likely that familiarity and identity are founded on the same or closely interrelated cues of configuration of the basic facial elements. Configural, because with 11 or 14 pixels/face quantization identity of local elements of face morphology is difficult to recognize.

We also examined effects of masking on pixelated facial image perception and the corresponding ERP signatures. Pixelated faces (eight alternatives for identification, pixelation at 6, 11, and 20 pixels/face) were presented briefly, followed by a backward mask. A novel masking method was used in that the mask was a negative image of

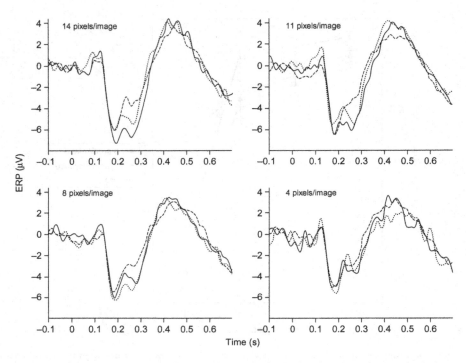

Figure 10.4 ERPs in response to pixelated images (shown at four levels of pixelation) as a function of image categories. Houses—green line (dashed line in print versions); unfamiliar faces—red line (dotted line in print versions); familiar faces—blue line (black line in print versions). Stimulus onset time at 0 s. ERPs do not differentiate image categories with the most-coarse pixelated image (4 pixels/face).

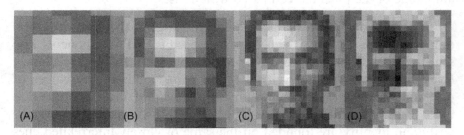

Figure 10.5 Examples of coarse-scale and intermediate-scale pixelated images (A–C) and a negative-image mask (D) used for masking the stimulus (C).

the preceding target, precisely overlapping in space. When superimposed simultaneously, positive- and negative-contrast pixel luminance levels counterbalance each other and all figural information is lost. With an increase in ISI targets gradually were released from this integrative zero-outcome masking. (ISI values were 0, 40, and 200 ms.) Examples of some typical stimuli images are depicted in Fig. 10.5.

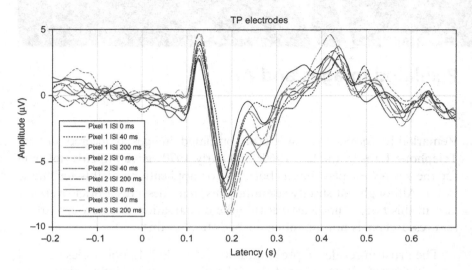

Figure 10.6 ERPs in response to pixelated and masked images of faces as a function of ISI between target and mask and level of pixelation. The amplitude of N170 is augmented when ISI >0 ms and level of pixelation is higher than 1 (ie, 11 or 20 pixels/face).

The amplitude of the face-specific ERP N170 became pronounced when ISI = 0 ms was changed to larger values and when the level of pixelation was equal to 11 or 20 pixels/face at least. (See also Fig. 10.6.) It should also be mentioned that with ISI = 0 ms behavioral results mostly indicated failure of face perception measured by objective and subjective (confidence ratings) measures, but ERP N170 was already present. This suggests that this method may be developed to become a test for preconscious processing of face-configuration information.

From the results of ERP experiments on pixelated facial image perception we can conclude that the main face-sensitive ERP component, N170, is largely a response to coarse-scale configuration of facial elements, not details of faces, and it can be brought about even when the fine-scale visual information extracted from the same spatial area is completely misleading and signaling different visual content. It is also necessary to mention that all pixelated images produced from the same original face, but spatially quantized over varying coarseness levels, have the same general space-average luminance and also largely overlapping SF content in the low-frequency bands. This means that distortion of the facial robust configuration must be the main reason why ERP amplitudes significantly change with change in coarseness level.

Pixelated Images and Art

Remarkably enough, the work on pixelated images pioneered in Bell Telephone Labs in the late 1960s and early 1970s gave impetus not only for the science of perception, but also for applications of this technique in art. Although not strictly scientific, this trend deserves a brief overview also in this book—not least because the demarcation line between artists and researchers is not unambiguously distinct in this case.

The artist-in-residence program initiated by Bell Laboratories in 1964 was informal, but nevertheless became highly influential. The development of computer hardware and software also allowed advances in computer graphics. Most importantly from the point of view of image pixelation, Ken Knowlton and Leon Harmon developed the computer-generated "Studies in Perception" series and together with the BEFLIX animation system (used to produce animated films), and in active inter-action with artists such as Stan VanDerBeek and Lillian Schwartz the science and technology advancements were adopted as new tools in visual art. (See also Noll, 2013; Schwartz, 1998.) In the mid 1960s Harmon suggested the principal technique of pixelation and Knowlton programmed it. In 1973 Leon Harmon wrote an article for *Scientific American* (Harmon, 1973), featuring the coarsely pixelated picture of Abraham Lincoln explaining that, when closely looked at, it initially appears to be just an array of abstract gray squares, but from a distant view the "block portrait" of the president could be perceived. A portrait of George Washington was not so well recognizable at the same coarseness level as that of Lincoln, presumably because of the very widely known and learned "standard five-dollar bill" portrait of Abraham Lincoln. Salvador Dali, who among other things was also fascinated by mosaic art, must have become aware of Harmon's work soon after and in 1974–76 in the process he incorporated the Lincoln picture into two versions of a painting—*Gala Contemplating the Mediterranean Sea which at 20 Meters becomes the Portrait of Abraham Lincoln (Homage to Rothko)* (oil on canvas, 1976; Salvador Dalí

Perception of Pixelated Images. DOI: http://dx.doi.org/10.1016/B978-0-12-809311-5.00011-8

Museum, St Petersburg, Florida; Museum in Figueras, Spain). In 1976 the picture was on exhibit at Guggenheim in New York.

Different from the Harmon's original style, Lincoln's face in Dali's painting is made up of pictures with full tonal ranges and articulated. The nude depicted in the picture took up several blocks and featured Dali's wife Gala. However, one of the local elements in the painting was Harmon's grayscale image of Lincoln. Dali deliberately made the blocks very large so that the edges of the blocks were not sharp and that for the large-scale Lincoln image viewing distance had to be large. If the observer is at a typical distance used for viewing art, (s)he is well aware that the picture is made up of multiple elements and the individual parts dominate, with the wholistic figure disappearing within the mass of smaller geometric forms. (This topic and a repro of the Dali's painting are nicely presented by Ramachandran and Rogers-Ramachandran, 2006.) Later, a limited edition (1240) lithograph based on the painting *Gala!* was created by Dali, labeled *Lincoln in Dalivision*. Nowadays these items, if authenticized, are sold with prices between $10,000 and $20,000. The principal art-historical significance of these Dali paintings is that they were among the first examples of a photomosaic approach in art and likely the most influential outside the artistic circles.

By a strange coincidence or something else, it was also in the 1970s that the American artist Chuck Close began creating his paintings reminiscent of pixel-based art. The visual appearance of these pictures was reminiscent of Impressionist/pointillist art or computer-generated images. Until late 1960s, Close had worked in the abstract expressionist style, but then—feeling that abstraction had lost its resources—he took a different direction, realism. Close founded his art on photographs, reproducing the reproduction of reality. He created large-scale portraits (often autoportraits), magnifying facial details, no matter how pleasing they were esthetically. The number of elements in his paintings varied from a few hundred to more than 100,000. The size of the pixels varied widely, reaching up to tens of centimeters. Usually a grid pattern was marked on a photograph and subsequently onto a canvas, maintaining the proportions. In order to transfer the image to the canvas, a set of coordinates was used. In many instances, and differently from the Harmon type of pixelation, Close places irregular forms within each pixel. This means that he paints an abstract "mark" in each pixel that differs in detail from the corresponding block of the photograph. The abstract local image and the corresponding local pixel of the original photo are conspicuously different, yet all of the local marks work together and create a realistic-looking portrait. From up close, the paintings appear flat and obscured, but from a far distance they seem to pop out and are experienced three-dimensionally. Here are some examples of his art:

- The 2.5 m high portrait of the artist Alex Katz made in 1987, consisting of 14,896 squares, each less than 2 cm;
- *Zhang Huan I*, 2008, oil on canvas, 257.8 cm × 213.4 cm (The Pace Gallery);
- *Emma*, 2002, 1 Chuck Close;
- *Self-Portrait*, 2000−01, oil on canvas, 274 cm × 213 cm. Art Supporting Foundation to the San Francisco Museum of Modern Art;
- *Bill II* (1991), oil on canvas, 92.4 cm × 76.2 cm.

It must be noted that the last painting from those listed above was psychophysically analyzed by Denis Pelli (1999). Studying Close's portrait paintings, he measured out that the transition from flat to three-dimensional perceptual impression occurs when the width of each local mark is 0.3 degrees of visual angle. However, Cavanagh and Kennedy (2000) argued that the smaller details within marks have also a significant effect.

To illustrate the value of the art by Chuck Close here is one example: a portrait that was first purchased for $9000 in 1972 sold for $4,832,000 at a Sotheby's auction in New York in May 2005. (For more detailed information about Close's biography and contributions the following sources can be recommended—McCarthy, 2005; Ravin & Odell, 2008.)

At this point an art-historical comment is necessary. Neither Knowlton and Harmon nor Chuck Close can be regarded as the very first who created images where the global Gestalt of the depiction is formed from local elements that do not necessarily share shape or form with the global entity. For example, Aboufadel, Boyenger, and Madsen, (2010, pp. 201−202) write: "The block portraits of Chuck Close are a special case of a method of creating an image known as a mosaic. Classically, a mosaic is an image created by combining small pieces of glass or stone of various colours to create a portrait or other subject, usually for a religious purpose. These tile pieces are known as tessera, and the mosaic is created by tiling or tessellating a region. ... mosaics exhibit the duality that one must either stand back from the image or squint one's eyes in order for the individual tessera to fuse together to see the portrait." Mosaics made from fragments or elements have been known since ancient times and, for example, in the case of Greek and Roman antiquity they were typically made of stone, metal, and glass. (Well-known examples of ancient mosaics come from Delos and Pompeii.) Their regularly spaced, usually colored elements are strikingly comparable with what contemporary computer art allows and also with Close's technique. Furthermore, impressionist artists—most notably the so-called pointillists—mastered the brush stroke so that an airy and apparently seamless impression could be produced by placing discrete dots on canvas. The standard example in this respect tends to be the painting *A Sunday Afternoon on the Island of Grand Jatte* by George Seurat from 1884 to 1886. When viewed from a close distance the multitude of small discrete dots are seen and the overall image is lost, but when observed from a more distant viewpoint, the dots cannot be discriminated, they assimilate and the picture shows a coherent scene full of meaning.

Also, an Austrian artist Gustav Klimt used discrete geometric elements in his many paintings where irregular shapes, usually rectangles, cover a large part of the canvas. Typically these are portraits, such as the portrait *Adele Bloch-Bauer I* of 1907. In the 1960s, pop artist Roy Lichtenstein capitalized on a technique common to printing technologies when half-tone impression is desired to be achieved just by applying discrete black dots of varying size and density. He magnified printed images, such as comic book illustrations, and was able to show the discrete dots inherent in commercial offset printing. Thus, the work of Leon Harmon coming from science and the work of Chuck Close coming from art had had a lot of precedence. On the other hand, prior to this work pixels with such a prominent size and different content from the wholistic image were not used.

What is obvious, however, is that after the 1970s, and related to the advancement of computing power and sophistication of software and image processing algorithms, computer art begun to get its momentum. The literature covering this domain is vast and we do not have space for this here. Let me simply refer to some characteristic work before the subtopic of art and image pixelation comes to close. First of all nonphotorealistic rendering (NPR) should be mentioned as a domain integrating image processing technology and art.

The paper by Battiato, Di Blasi, Farinella, and Gallo (2007) is a good source for this purpose. These authors note that NPR is a well-advanced area of computer graphics and is currently applied to several relevant contexts such as scientific visualization, information visualization, and *artistic style emulation*. Current NPR strives to digitally reproduce artistic media (eg, watercolors, crayons, charcoal) and artistic styles (eg, cubism, impressionism, pointillism). NPR includes subdisciplines, such as artistic rendering and computational aesthetics, among others. Battiato and colleagues define this generic domain through its main purpose as follows: *to reproduce the aesthetic essence of arts by mean of computational tools*. Among the varied styles and methods, the problem of digital mosaic creation stands as the one perhaps closest to our pixelation topic. In this "mosaic domain," techniques have been developed that make use of primitives larger than elementary system pixels, points, or lines. Borrowed from older use, the terms "tiles" and "tessellation" have now acquired their more modern connotation. Thus, in the digital realm, "mosaics are illustrations composed by a collection of small images called 'tiles' and the tiles tessellate a source image with the purpose of reproducing the original visual information rendered into a new mosaic-like style" (Battiato et al., 2007, p. 794).

Four different mosaic types were listed: (1) crystallization mosaics (ie, tessellation); (2) ancient mosaics; (3) photomosaics; and (4) puzzle image mosaics (Battiato et al., 2007). For our purposes, the photomosaic is closest to the common psychophysical pixelation domain, as well as the method of Chuck Close, and is also one of the best-developed fields of digital mosaics. It transforms an input image into a rectangular grid of thumbnail images, with its basic algorithm searching a large database of images for one that approximates a block of pixels in the main image. While in Harmon (1973) and Harmon and Julesz (1973) the "tiles" were blocks or pixels with internally uniform brightness and color, photomosaics appear more complex and become closer to visual art (also not least because of the virtually endless variants of how to substantiate the individual blocks/tiles).

Battiato and colleagues give a handful of fine examples of digital mosaic art, including their own (eg, by Di Blasi and colleagues). In related work Battiato, Di Blasi, Farinella, and Gallo (2006) and Di Blasi, Gallo, and Petralia (2006) provide a useful overview of digital mosaic production techniques and approaches. For example, in the domain of crystallization mosaics, by using computational geometry (Voronoi diagrams) together with image processing, these techniques lead to mosaics that simulate the typical effect of some glass windows in churches. (Among others, approaches by Haeberli and Dobshi et al. are referred to.) Aboufadel et al. (2010) describe a novel method—using wavelet filters and a careful analysis of color—to digitally imitate Chuck Close's block portraits. Furthermore, tilings other than squares can be used, such as diagonal tiling and triangle tiling. And, of course, some of it was difficult to demonstrate without using the celebrated portrait of Abraham Lincoln.

CHAPTER 12

What the Results of Pixelation-Based Studies Have Told Us About the Nature of Perception

For obvious reasons, pixelated images as an experimental aid are used first of all in the domain of pattern or form perception. According to Uttal (1994), pattern or form recognition is the process by which visually presented objects (images of objects) are identified, categorized, and named. By performing operations of pattern-information processing the visual system basically attempts to answer the question, what it is that is being seen. In this process there can be constraints put upon the processing operations—spatial, temporal, contrast-related, noise-related, context-related, attention-related, learning-related, and subject's state-related constraints. If researchers wittingly and purposely use the constraints as independent variables, processing regularities and the nature of the mechanisms of perception can be revealed.

Employing pixelated images for perception research touches the very essence of the central methodological means of perception research by virtue of introducing quantifiable and precisely measurable changes into stimulus image, with possibilities of meaningful interpretation of the effects pixelation causes depending on the variation in the parameters of pixelation. At this point let me present an essential comment on the usefulness of pixelation as the experimental technique. Traditionally, other means to manipulate spatial stimuli characteristics have been used—spatial frequency (SF) Fourier filtering, Gaussian filtering, adding parametrically dosed out visual noise, and some other manipulations. However, each of these methods has certain limitations. Using random noise brings in uncertainty and this method also does not manipulate with the stimulus-image itself in a meaningful way. Traditional filtering approaches are free from these shortcomings, but they are weak in terms of their capacity to manipulate or distort configuration in the way susceptible for precise measureable specification. Also, in most cases the varieties of straightforward SF filtering do not introduce variability of image interpretation whereby the

Perception of Pixelated Images. DOI: http://dx.doi.org/10.1016/B978-0-12-809311-5.00012-X

representation is categorized or named in one or another way. Pixelation has all these virtues. For example, why the traditional SF-filtering approach may mislead researchers who are interested in the nature of the mechanisms and processes of object image recognition? If we presume that low-level vision is proficient in detecting the maxima and minima of local contrast or brightness then it is easy to show that low-pass filtered images carry virtual amodal contours that can be represented as pseudo-high-frequency, shape-defining lines (shapes from lines). One way to do this is to execute an imaginary connection of isoluminance local points (but this need not be the only viable routine). This, in turn, means that despite the filtering out of real high SF (HSF) components, virtual HSF-like cues of configuration are not eliminated. Therefore, the methods that both (1) filter out the HSF content from the original image and (2) create distortions and/or uncertainty in the localization of the local contrast maxima and minima may be a better experimental solution for several research objectives. This refers especially to the domain of studies where topics of configuration and coarseness of spatial representation stand at the centerstage. Bearing this in mind, let me now discuss the contribution of research capitalizing on image pixelation for the basic issues of perception research.

What are the central problems to be solved or at least made more manageable in perception research that the paradigm of pixelation could help resolve? From the several items of the theoretical agenda typical for pattern/form recognition research pointed out by Uttal (1994) there are some that are especially relevant for the pixelation paradigm. First, there is the issue of *elementalism versus holism*: according to the first of these perspectives, parts (elements, features) of an object image are the main "trustees" for recognition and also tend to have precedence in image processing; according to the holistic view the entire object or the arrangement of its parts is decisive for recognition and dominates processing. A subservant problem relates to the time course and temporal order of processing elements versus wholistic aspects of the stimulus object. Second, the problem of *stimulus equivalence* has been acknowledged as one of the central issues in perception research: how is it that human visual system is able to recognize objects despite the substantial variability in viewing conditions and object states such as size, viewpoint, or orientation changes, dynamic transformations, inclusion in noise, etc. We can add to this set of problems also the issue of the mode and extent of top-down processing applied on representations built up during the initial feed-forward stage of image processing by the brain in a bottom-up manner. Related to this, there is the issue of *task dependency* of processing the same objects which primarily speaks to the top-down effects.

1. *Elementalism versus holism*. The majority of the studies capitalizing on the image pixelation technique and that were described and analyzed in the preceding chapters showed that in most cases the holistic, Gestalt-tradition-related view can be accepted. Indeed, when specific task demands are used (including some specific manipulations with the diagnosticity of a particular level of spatial scale) or discrimination is made difficult, elements and local features acquire a more important role in identification and recognition. However, the simple fact that coarse-scale pixelated images (eg, with less than 20 pix/image resolution) allow quite high-level identification or recognition speaks for itself. From the time-course point of view of what regards the order of processing different spatial scales, most of the research with pixelated images supports the coarse-to-fine regularity of processing. Bearing in mind that holistic versus elemental, relational (configural) versus featural, coarse versus fine, and global versus local dichotomies are not equivalent (Kimchi, 1992; Morrison & Schyns, 2001; Piepers & Robbins, 2012) and accepting that task-dependent scale diagnosticity may influence (and in some cases even reverse) processing order (Hegdé, 2008; Morrison & Schyns, 2001), pixelation-based research in general supports the coarse-to-fine routine. This is consistent with the views of the majority of authors, mostly founded on the empirical results from psychophysics and neuroscience (Bachmann, 1991; Gao & Bentin, 2011; Goffaux et al., 2011; Loftus & Harley, 2004; Meinhardt-Injac, Persike, & Meinhardt, 2010; Mermillod, Guyader, & Chauvin, 2005; Morgan & Watt, 1997; Musel et al., 2014; Neri, 2011; Otsuka et al., 2014; Parker et al., 1997; Sugase, Yamane, Ueno, & Kawano, 1999).

2. *Stimulus equivalence*. By introducing the gradually more coarse level of pixelation, the equivalence of the original source image and the pixelated version also becomes gradually more perturbed, making it progressively more difficult for the visual system to find the correct identity label or category for the perceived pixelated image. With fine-scale pixelation, the visual system carries out basically the same

routines as with the original image because coarse-scale configural information and local feature cues in both instances signal a veridical set of information sources for successful processing. At a certain point along the coarseness scale value (when it is coarsened) equivalence establishment becomes impossible or ambiguous. In the tasks requiring identification as supported by actual or memory-based alternatives it is especially the distortion of configuration by the pixel-blocks which makes an obstacle for one-to-one correspondence verification between the original and the pixelized versions. Additionally, image interpretation in terms of an unknown mosaic-object instead of the veridical visual categorization of the unquantized original is another source of misleading the processing operations necessary for equivalence establishment. Pixelation-based studies showed that there is a critical range of pixelation values above which the pixelation transform is tolerated and veridical correspondence between the pixelated version and original is possible. As local image features are dissolved when intermediate or coarse-scale pixelation is used, the equivalence task is internally solved by some configural processing operations. In many identification and recognition tasks this critical value of pixelation remains between about 10 and 20 pix/image. The critical values in matching or comparison tasks tend to be less predictable; this is probably because of the other usable cues besides coarse-scale-based configuration information. In many cases equivalence can be established based on coarse-scale information and in the presence of the masking and misleading cues from the mosaic of pixels.

3. *Task dependence.* The pixelation-based research reviewed in the preceding chapters indicated that the effects of certain particular levels of pixelation as critical or the effects of combined degradations considerably depend on the task. Matching and comparison tasks can in many occasions be successfully carried out with more coarse pixelation levels than identification tasks. However, when the set of items to be compared is mutually highly similar, matching also becomes difficult. However, in general, with many different tasks there is a surprising tolerance of the perceptual operations to a quite coarse level of pixelation. Person or object identification, sequential matching, celebrity recognition, emotion discrimination, appearance-based trait perception can all be completed above chance and sometimes very well with coarse-scale pixelated images. Additionally, the pixelated image-based perception is susceptible to attentional effects, including the

paradoxical decrease in accuracy with preset focused attention. There was also some indication that—as indicated by event-related potentials—the brain may sometimes discriminate pixelated object categories at the same time when the observer behavioral responses are not veridical.

I would like to hold that the paradigm of pixelation very well suits the theoretical school of perceptual microgenesis (for review, see Bachmann, 2000). While up to now the majority of perception research using the pixelation technique has been concentrated on face perception, in principle this approach can be extended even more to other complex stimuli (eg, notice the comments and suggestions by Carbon, 2013). As pointed out also by Carbon (2013), by combining image degradations like pixelation with presentation time-limitation techniques, further insights into the processing of complex visual stimuli may be attained. Thus, by combining the presentation of systematically degraded stimuli at different processing times, specific characteristics of the stimulus object could be identified as relevant at different processing stages. This allows to assess specific stimulus qualities involved in the microgenetic process at its different stages and as related to different tasks. The advantageous feature of the pixelation approach when combined with microgenetic temporal manipulations consists of its natural appropriateness for manipulating and precisely dosing out the variables of the two fundamental attributes of reality— space and time. Microgenetic research on complex image perception need not be limited to the traditional psychophysics, but can also be set to engage art perception studies (Augustin, Defranceschi, Fuchs, Carbon, & Hutzler, 2011).

CHAPTER 13

Conclusions

Now, this brings us to the end of the presented succinct survey of the perception research carried out in the paradigm of pixelated images. We have seen that this approach has its definite merits:

- Pixelation transform is straightforward and simple in principle, but meaningfully related to several basic variables essential for perception research.
- The variable values of the level of pixelation can be precisely measured and changed.
- Pixelation transform can be meaningfully related to several staple procedures and measures used in perception research, such as spatial frequency (SF) spectra, luminance contrast, gray levels, image size, filtering operations, interelement relations, etc.
- This paradigm is meaningfully and closely related to several other influential perception-research paradigms—perception as SF processing, global versus local perception, features versus configuration (or part vs relation) processing, perceptual organization (eg, Gestalt approach), microgenetic studies, etc.
- Pixelation allows control over variables that cannot be parametrically controlled by some traditional methods of stimulus-image transformation.
- This technique is useful for research on the problems where feature versus configuration, coarse scale versus fine scale, and invariantly interpretable versus multiply interpretable experimental contrasts must be tested.
- The pixelation paradigm has merit theoretically, as well as from the applied point of view, because pixelated images are frequently used in media, other communication systems, and other IT applications—mass communication, forensic context, bitrate optimization, and computer art come to mind as the first examples.

Perception of Pixelated Images. DOI: http://dx.doi.org/10.1016/B978-0-12-809311-5.00013-1

There are also some common misinterpretations of what regards the visual nature of the pixelation transform and its perceptual effects. Pixelation is often regarded as another means of SF filtering and it is believed that traditional SF filtering and pixelation can be used for the same purposes bearing the same effects. This is wrong, first of all because in addition to SF filtering, pixelation also has a more severe effect on configuration of the original image and by virtue of the mosaic structure and its sharp edges and corners, a pixelated image is a combination of the remaining (mostly lower SF) attributes of the original image, masking cues, and also an alternative object-category interpretation. In the common practice of using pixelation for achieving anonymity of the depicted persons, the practitioners often forget or they did not know that there are many procedures available and accidentally occurring conditions which likely make the supposedly disguised face recognizable.

Thus, in a sense, a pixelated image, when a coarse-scale level of pixelation is used, can be classified as a multiply interpretable, ambiguous object allowing interpretation equivalent to the original, prepixelation stage image as well as a collection of blocks or a kind of mosaic pattern different in form and meaning from the—now hidden—original category or identity.

It is typical for most of the tasks with pixelated images as stimuli that a monotonic change of the level of coarseness of pixelation does not lead to a proportional change in the dependent perceptual effects, especially the accuracy of perception. Usually there is a certain critical level of pixelation beyond which an abrupt step-like decrease in the effect occurs. Pixelation values and the values of the effects brought about by pixelation do not have a linear relationship. Related to this, the optima and limiting values of pixelation can be established for investigative and practical purposes.

Specific values of the level of pixelation have their meaning mostly in terms of the number of pixels per object image, not in terms of their absolute values (which is similar to most of the SF research in what regards accuracy and efficiency of processing). There are, of course, certain limiting conditions for this regularity, first of all related to the visual acuity thresholds.

What may seem surprising, is that there are many different visual tasks and information-processing domains where the tolerance of the engaged perceptual processing system to the quite coarse level of pixelation has become evident. This list includes: facial identification, personal familiarity perception, perception of appearance-based personality-trait cues, perceiving visible speech cues, and extracting attractiveness cues from pixelated images.

Similarly to some other, related research, pixelation-assisted studies have supported several theoretical stances of the science of perception, among which the main ones are the following: perception is a microgenetic event unfolding in real time over successive stages, each of which has different contents; in most cases perception unfolds from coarse-scale to fine-scale representation; in many cases configural relations between the parts of an object image are the prime source of veridical or task-complying processing compared to parts-based processing; perception is a flexible process where the same stimulus-object can be interpreted, in principle, in alternative ways.

Last, but not least, I do hope that when you stare at a pixelated image of a handsome object quantized, say, at the level of 11 pixels/ image along its one axis, you agree that there is some certain aesthetic appeal in this picture. Perhaps this observation helps to guess what was the possible intuitive motivation for why the present author has spent quite so much time researching the mosaic images and their effects.

REFERENCES

Aboufadel, E., Boyenger, S., & Madsen, C. (2010). Digital creation of Chuck Close block-style portraits using wavelet filters. *Journal of Mathematics and the Arts*, *4*(4), 201–211.

Aguado, L., Serrano-Pedraza, I., Rodriguez, S., & Román, F. J. (2010). Effects of spatial frequency content on classification of face gender and expression. *The Spanish Journal of Psychology*, *13*, 525–537.

Albanesi, M. G., & Amadeo, R. (2014). A new categorization of image quality metrics based on a model of human quality perception. *International Scholarly and Scientific Research & Innovation*, *8*(6), 921–929.

Allison, T. H., Ginter, G., McCarthy, A. C., Nobre, A., Puce, M., Luby, D., et al. (1994). Face recognition in human extrastriate cortex. *Journal of Neurophysiology*, *71*, 821–825.

Amishav, R., & Kimchi, R. (2010). Perceptual integrality of componential and configural information in faces. *Psychonomic Bulletin & Review*, *17*, 743–748.

Andres, A. J. D., & Fernandes, M. A. (2006). Effect of short and long exposure duration and dual-tasking on a global-local task. *Acta Psychologica*, *122*, 247–266.

Antes, J. R., & Mann, S. W. (1984). Global-local precedence in picture processing. *Psychological Research*, *46*, 247–259.

Antes, J. R., Penland, J. G., & Metzger, R. L. (1981). Processing global information in briefly presented pictures. *Psychological Research*, *43*, 277–292.

Augustin, M. D., Defranceschi, B., Fuchs, H. K., Carbon, C.-C., & Hutzler, F. (2011). The neural time course of art perception: An ERP study on the processing of style versus content in art. *Neuropsychologia*, *49*, 2071–2081.

Awasthi, B., Friedman, J., & Williams, M. A. (2011). Faster, stronger, lateralized: Low spatial frequency information supports face processing. *Neuropsychologia*, *49*, 3583–3590.

Awasthi, B., Sowman, P. F., Friedman, J., & Williams, M. A. (2013). Distinct spatial scale sensitivities for early categorization of faces and places: Neuromagnetic and behavioral findings. *Frontiers in Human Neuroscience*, *7*. Available from http://dx.doi.org/10.3389/fnhum.2013.00091.

Bachmann, T. (1980). Genesis of a subjective image. *Acta et Commentationes Universitatis Tartuensis. #522. Problems of Cognitive Psychology*, 102–126.

Bachmann, T. (1987). Different trends in perceptual pattern microgenesis as a function of the spatial range of local brightness averaging. *Psychological Research*, *49*, 107–111.

Bachmann, T. (1989). Microgenesis as traced by the transient paired-forms paradigm. *Acta Psychologica*, *70*, 3–17.

Bachmann, T. (1991). Identification of spatially quantised tachistoscopic images of faces: How many pixels does it take to carry identity? *European Journal of Cognitive Psychology*, *3*, 87–103.

Bachmann, T. (1994). *Psychophysiology of visual masking. The fine structure of conscious experience*. Commack, NY: Nova Science Publishers, Inc.

Bachmann, T. (2000). *Microgenetic approach to the conscious mind*. Amsterdam/Philadelphia, PA: John Benjamins.

Bachmann, T. (2007). When beauty breaks down: Investigation of the effect of spatial quantisation on aesthetic evaluation of facial images. *Perception*, *36*, 840–849.

Bachmann, T., & Allik, J. (1976). Integration and interruption in the masking of form by form. *Perception, 5,* 79–97.

Bachmann, T., & Kahusk, N. (1997). The effects of coarseness of quantisation, exposure duration, and selective spatial attention on the perception of spatially quantised ("blocked") visual images. *Perception, 26,* 1181–1196.

Bachmann, T., & Leigh-Pemberton, L. (2002). Binocular rivalry as a function of spatial quantisation of the images of faces: Precategorical level controls it. *Trames, 6*(4), 283–296.

Bachmann, T., Luiga, I., & Põder, E. (2004). Forward masking of faces by spatially quantized random and structured masks. *Psychological Research/Psychologische Forschung, 69,* 11–21.

Bachmann, T., Luiga, I., & Põder, E. (2005a). Variations in backward masking with different masking stimuli: I. Local interaction versus attentional switch. *Perception, 34,* 131–137.

Bachmann, T., Luiga, I., & Põder, E. (2005b). Variations in backward masking with different masking stimuli: II. The effects of spatially quantised masks in the light of local contour interaction, interchannel inhibition, perceptual retouch, and substitution theories. *Perception, 34,* 139–154.

Bardi, L., Kanai, R., Mapelli, D., & Walsh, V. (2012). Direct current stimulation (tDCS) reveals parietal asymmetry in local/global and salience-based selection. *Cortex, 49,* 850–860.

Baron, W. S., & Westheimer, G. (1973). Visual acuity as a function of exposure duration. *Journal of the Optical Society of America, 63,* 212–219.

Battiato, S., Di Blasi, G., Farinella, G. M., & Gallo, G. (2006). A survey of digital mosaic techniques. In G. Gallo, S. Battiato, & F. Stanco (Eds.), *Eurographics italian chapter conference (2006).*

Battiato, S., Di Blasi, G., Farinella, G. M., & Gallo, G. (2007). Digital mosaic frameworks—An overview. *Computer Graphics Forum, 26,* 794–812.

Beaucousin, V., Cassotti, M., Simon, G., Pineau, A., Kostova, M., Houdé, O., et al. (2011). ERP evidence of a meaningfulness impact on visual global/local processing: When meaning captures atgtention. *Neuropsychologia, 49,* 1258–1266.

Bentin, S., Allison, T., Puce, A., Perez, E., & McCarthy, G. (1996). Electrophysiological studies of face perception in humans. *Journal of Cognitive Neuroscience, 8,* 551–565.

Berry, D. S., Kean, K. J., Misovich, S. J., & Baron, R. M. (1991). Quantized displays of human movement: A methodological alternative to the point-light display. *Journal of Nonverbal Behavior, 15,* 81–97.

Bex, P. J., Solomon, S. G., & Dakin, S. C. (2009). Contrast sensitivity in natural scenes depends on edge as well as spatial frequency structure. *Journal of Vision, 9*(10), 1 ,1–19, http://journalofvision.org/9/10/1/, http://dx.doi.org/10.1167/9.10.1

Bhatia, S. K., Lakshminarayanan, V., Samal, A., & Welland, G. V. (1995). Human face perception in degraded images. *Journal of Visual Communication and Image Representation, 6,* 280–295.

Biederman, I., & Kalocsai, P. (1997). Neurocomputational bases of object and face recognition. *Philosophical Transactions of the Royal Society of London. B, 352,* 1203–1219.

Bindemann, M., Attard, J., Leach, A., & Johnston, R. A. (2013). The effect of image pixelation on unfamiliar-face matching. *Applied Cognitive Psychology, 27,* 707–717.

Boer, L. C., & Keuss, P. J. G. (1982). Global precedence as a postperceptual effect: An analysis of speed-accuracy trade-off functions. *Perception and Psychophysics, 31,* 358–366.

Borgo, R., Proctor, K., Chen, M., Jänicke, H., Murray, T., & Thornton, I. M. (2010). Evaluating the impact of task demands and block resolution on the effectiveness of pixel-based visualization. *IEEE Transactions on Visualization and Computer Graphics, 16,* 963–972.

Brady, N. (1997). Spatial scale interactions and image statistics. *Perception, 26,* 1089–1100.

Brand, J., & Johnson, A. P. (2014). Attention to local and global levels of hierarchical Navon figures affects rapid scene categorization. *Frontiers in Psychology, 5*, 1274. Available from http://dx.doi.org/10.3389/fpsyg.2014.01274.

Breese, B. B. (1899). On inhibition. *Psychological Review Monograph, 3*, Whole No. 2

Breitmeyer, B. G. (1984). *Visual masking: An integrative approach.* Oxford: Clarendon.

Breitmeyer, B. G., & Öğmen, H. (2000). Recent models and findings in backward masking: A comparison, review and update. *Perception and Psychophysics, 62*, 1572−1595.

Breitmeyer, B. G., & Öğmen, H. (2006). *Visual masking.* Oxford: Oxford University Press.

Brooke, N. M., & Templeton, P. D. (1990). Visual speech intelligibility of digitally processed facial images. *Proceedings of the Institute of Acoustics, 12*, 483−490.

Bruce, V., & Young, A. (1998). *In the eye of the beholder: The science of face perception.* Oxford: Oxford University Press.

Bruyer, R. (2011). Configural face processing: A meta-analytic review. *Perception, 40*, 1478−1490.

Bukach, C. M., Gauthier, I., & Tarr, M. J. (2006). Beyond faces and modularity: The power of an expertise framework. *Trends in Cognitive Sciences, 10*, 159−166.

Burger, W., & Burge, M. J. (2013). *Principles of digital image processing.* Berlin: Springer- Verlag.

Burton, A. M., Schweinberger, S. R., Jenkins, R., & Kaufmann, J. M. (2015). Arguments against a configural processing account of familiar face recognition. *Perspectives on Psychological Science, 10*(4), 482−496.

Burton, A. M., Wilson, S., Cowan, M., & Bruce, V. (1999). Face recognition in poor-quality video: Evidence from security surveillance. *Psychological Science, 10*, 243−248.

Busey, T. A., & Loftus, G. R. (2007). Cognitive science and the law. *Trends in Cognitive Sciences, 11*, 111−117.

Caelli, T., & Yuzyk, J. (1985). What is perceived when two images are combined? *Perception, 14*, 41−48.

Caharel, S., Ramon, M., & Rossion, B. (2014). Face familiarity decisions take 200 msec in the human brain: Electrophysiological evidence from a go/no-go speeded task. *Journal of Cognitive Neuroscience, 26*, 81−95.

Calis, G., Sterenborg, J., & Maarse, F. (1984). Initial microgenetic steps in single-glance face recognition. *Acta Psychologica, 55*, 215−230.

Campbell, C. S., & Massaro, D. W. (1997). Perception of visible speech: Influence of spatial quantization. *Perception, 26*, 129−146.

Caplette, L., West, G., Gomot, M., Gosselin, F., & Wicker, B. (2014). Affective and contextual values modulate spatial frequency use in object recognition. *Frontiers in Psychology, 5*, 512. Available from http://dx.doi.org/10.3389/fpsyg.2014.00512.

Carbon, C.-C. (2013). Creating a framework for experimentally testing early visual processing: A response to Nurmoja, et al. (2012) on trait perception from pixelized faces. *Perceptual and Motor Skills, 117*, 1−4.

Carretié, L., Rios, M., Perianes, J. A., Kessel, D., & Álvares-Linera, J. (2012). The role of low and high spatial frequencies in exogenous attention to biologically salient stimuli. *PLoS ONE, 7* (5), e37082. Available from http://dx.doi.org/10.1371/journal.pone.0037082.

Cavanagh, P., & Kennedy, J. (2000). Close encounters: Details veto depth from shadows. *Science, 287*, 2423−2425.

Chalupa, L. M., & Werner, J. S. (2004). *The visual neurosciences.* Cambridge, MA: MIT Press.

Chaudhry, S., & Singh, K. (2012). A brief introduction of digital image processing. *International Journal of Advances in Computing and Information Technology.* Available from http://dx.doi.org/ 10.6088/ijacit.12.10010.

Chetouani, A., Deriche, M., & Beghdadi, A. (2010). *Classification of image distortions using image quality metrics and linear discriminant analysis. 18th European signal processing conference (EUSIPCO-2010), Aalborg, Denmark, August 23–27* (pp. 319–322).

Christie, J., Ginsberg, J. P., Steedman, J., Fridriksson, J., Bonilha, L., & Rorden, C. (2012). Global versus local processing: Seeing the left side of the forest and the right side of the trees. *Frontiers in Human Neuroscience, 6*, 28. Available from http://dx.doi.org/10.3389/fnhum.2012.00028.

Collin, C. A., Liu, C. H., Troje, N. F., McCullen, P. A., & Chaudhuri, A. (2004). Face recognition is affected by similarity in spatial frequency range to a greater degree than within-category object recognition. *Journal of Experimental Psychology: Human Perception and Performance, 30*, 975–987.

Collin, C. A., Rainville, S., Watier, N., & Boutet, I. (2014). Configural and featural discriminations use the same spatial frequencies: A model observer versus human observer analysis. *Perception, 43*(6), 509–526.

Conci, M., Töllner, T., Leszczynski, M., & Müller, H. J. (2011). The time-course of global and local attentional guidance in Kanizsa-figure. *Neuropsychologia, 49*, 2456–2464.

Coren, S., Ward, L. M., & Enns, J. T. (1999). *Sensation and perception.* London: Harcourt Brace.

Costen, N. P., Parker, D. M., & Craw, I. (1994). Spatial content and spatial quantisation effects in face recognition. *Perception, 23*, 129–146.

Costen, N. P., Parker, D. M., & Craw, I. (1996). Effects of high-pass and low-pass spatial filtering on face identification. *Perception and Psychophysics, 58*, 602–612.

Costen, N. P., Shepherd, J. W., Ellis, H. D., & Craw, I. (1994). Masking of faces by facial and non-facial stimuli. *Visual Cognition, 1*, 227–251.

Curby, K. M., Goldstein, R. R., & Blacker, K. (2013). Disrupting perceptual grouping of face parts impairs holistic face processing. *Attention, Perception, & Psychophysics, 75*, 83–91.

Dale, G., & Arnell, K. M. (2013). Investigating the stability and relationships among global/local processing measures. *Attention, Perception, & Psychophysics, 75*, 394–406.

Davies, G., Ellis, H., & Shepherd, J. (Eds.), (1981). *Perceiving and remembering faces* London: Academic Press.

Deangelis, G. C., & Anzai, A. (2004). A modern view of the classical receptive field: Linear and nonlinear spatiotemporal processing by V1 neurons. In L. M. Chalupa, & J. S. Werner (Eds.), *The visual neurosciences* (pp. 704–719). Cambridge, MA: MIT Press.

De Cesarei, A., & Loftus, G. R. (2011). Global and local vision in natural scene identification. *Psychonomic Bulletin & Review, 18*, 840–847.

Deco, G., & Heinke, D. (2007). Attention and spatial resolution: A theoretical and experimental study of visual search in hierarchical pattterns. *Perception, 36*, 335–354.

De Gelder, B., & Rouw, R. (2001). Beyond localisation: A dynamical dual route account of face recognition. *Acta Psychologica, 107*, 183–207.

Demanet, J., Dhont, K., Notebaert, L., Pattyn, S., & Vandierendonck, A. (2007). Pixelating familiar people in the media: Should masking be taken at face value? *Psychologica Belgica, 47*, 261–276.

Diamond, R., & Carey, S. (1986). Why faces are not special: An effect of expertise. *Journal of Experimental Psychology: General, 115*, 107–117.

Di Blasi, G., Gallo, G., & Petralia, M. P. (2006). Smart ideas for photomosaic rendering. In G. Gallo, S. Battiato, & F. Stanco (Eds.), *Eurographics italian chapter conference (2006).*

Dosselmann, R., & Yang, X. D. (2013). *A rank-order comparison of image quality metrics. 26th IEEE canadian conference of electrical and computer engineering (CCECE)* (pp. 1−4).

Dulaney, C. L., & Marks, W. (2007). The effects of training and transfer on global/local processing. *Acta Psychologica, 125*, 203−220.

Durgin, F. H., & Profitt, D. R. (1993). *Perceptual response to visual noise and display media. Final Report, NASA-NAG2-814.* University of Virginia.

Ehrlich, S. M., Schiano, D. J., & Sheridan, K. (2000). *Communicating facial affect: It's not the realism, it's the motion. Proceedings of ACM CHI 2000 conference on human factors in computing systems* (pp. 252−253). NY: ACM.

Eimer, M. (2011). The face-sensitive N170 component of the event-related brain potential. In A. J. Calder, G. Rhodes, M. Johnson, & J. Haxby (Eds.), *The oxford handbook of face perception.* Oxford: Oxford University Press.

Ekman, P., & Friesen, W. V. (1976). *Pictures of facial affect.* Palo-Alto, CA: Consulting Psychologists Press.

Endo, M., Kirita, T., & Abe, T. (1995). The effects of image blurring on the recognition of facial expressions. *Tohoku Psychologica Folia, 54*, 68−82.

Eriksen, C. W., & StJames, J. D. (1986). Visual attention within and around the field of focal attention: A zoom lens model. *Perception and Psychophysics, 40*, 225−240.

Field, D. J. (1999). Wavelets, vision and the statistics of natural scenes. *Philosophical Transactions of the Royal Society of London A, 357*, 2527−2542.

Fink, G. R., Halligan, P. V., Marshall, J. C., Frith, C. D., Frackowiak, R. S. J., & Dolan, R. J. (1996). Where in the brain does visual attention select the forest and the trees? *Nature, 382*, 626−628.

Fiorentini, A., Maffei, L., & Sandini, G. (1983). The role of high spatial frequencies in face perception. *Perception, 12*, 195−201.

Flevaris, A. V., Bentin, S., & Robertson, L. C. (2011a). Attention to hierarchical level influences attentional selection of spatial scale. *Journal of Experimental Psychology: Human Perception and Performance, 37*, 12−22.

Flevaris, A. V., Bentin, S., & Robertson, L. C. (2011b). Attentional selection of relative SF mediates global versus local processing: Evidence from EEG. *Journal of Vision, 11*(7), 1−12 , 11

Flevaris, A. V., Robertson, L. C., & Bentin, S. (2008). Using spatial frequency scales for processing face features and face configuration: An ERP analysis. *Brain Research, 1194*, 100−109.

Förster, J. (2012). GLOMO[sys]: The how and why of global and local processing. *Current Directions in Psychological Science, 21*, 15−19.

Frowd, C. D., et al. (2005). A forensically valid comparison of facial composite systems. *Psychology, Crime and Law, 11*, 33−52.

Gable, P. A., Poole, B. D., & Cook, M. S. (2013). Asymmatrical hemisphere activation enhances global-local processing. *Brain and Cognition, 83*, 337−341.

Gao, Z., & Bentin, S. (2011). Coarse-to-fine encoding of spatial frequency information into visual short-term memory for faces but impartial decay. *Journal of Experimental Psychology: Human Perception and Performance, 37*, 1051−1064.

Gerlach, C., & Krumborg, J. R. (2014). Same, same—but different: On the use of Navon derived measures of global/local processing in studies of face processing. *Acta Psychologica, 153*, 28−38.

Gerstner, T. D., De Carlo, D., Alexa, M., Finkelstein, A., Gingold, Y., & Nealen, A. (2012). *Pixelated image abstraction. Proceedings of the symposium on non-photorealistic animation and rendering, NPAR '12* (pp. 29−36). Aire-la-Ville, Switzerland: Eurographics Association.

Gilad-Gutnick, S., Yovel, G., & Sinha, P. (2012). Recognizing degraded faces: The contribution of configural and featural cues. *Perception, 41*, 1497–1511.

Goffaux, V. (2012). The discriminability of local cues determines the strength of holistic face processing. *Vision Research, 64*, 17–22.

Goffaux, V., Gauthier, I., & Rossion, B. (2003). Spatial scale contribution to early visual differences between face and object processing. *Cognitive Brain Research, 16*, 416–424.

Goffaux, V., Hault, B., Michel, C., Vuong, Q. C., & Rossion, B. (2005). The respective role of low and high spatial frequencies in supporting configural and featural processing of faces. *Perception, 34*, 77–86.

Goffaux, V., Jemel, B., Jacques, C., Rossion, B., & Schyns, P. G. (2003). ERP evidence for task modulations on face perceptual processing at different spatial scales. *Cognitive Science, 27*, 313–325.

Goffaux, V., Peters, J., Haubrechts, J., Schiltz, C., Jansma, B., & Goebel, R. (2011). From coarse to fine? Spatial and temporal dynamics of cortical face processing. *Cerebral Cortex, 21*, 467–476.

Gold, J., Bennett, P. J., & Sekuler, A. B. (1999). Identification of band-pass filtered letters and faces by human and ideal observers. *Vision Research, 39*, 3537–3560.

Gonzalez, R. C., & Woods, R. E. (2008). *Digital image processing*. Upper Saddle River, NJ: Pearson Prentice Hall.

Gore, A., & Gupta, S. (2015). Full reference image quality metrics for JPEG compressed images. *International Journal of Electronics and Communications (AEÜ), 69*, 604–608.

Gosselin, F., & Schyns, P. G. (2002). *RAP*: A new framework for visual categorization. *Trends in Cognitive Sciences, 6*, 70–77.

Grice, G. R., Canham, L., & Boroughs, J. M. (1983). Forest before trees? It depends where you look. *Perception and Psychophysics, 33*, 121–128.

Halit, H., de Haan, M., Schyns, P. G., & Johnson, M. H. (2006). Is high-spatial frequency information used in the early stages of face detection?. *Brain Research, 1117*, 154–161.

Hallum, L.E. (2007). *Prosthetic vision: Visual modelling, information theory and neural correlates*, May 2007, thesis paper submitted in partial fulfilment of the requirements of the degree of Doctor of Philosophy, Graduate School of Biomedical Engineering, University of New South Wales.

Han, S., Fan, S., Chen, L., & Zhuo, Y. (1997). On the different processing of wholes and parts: A psychophysiological analysis. *Journal of Cognitive Neuroscience, 9*, 685–698.

Hanso, L., Murd, C., & Bachmann, T. (2010). Tolerance of the ERP signatures of unfamiliar versus familiar face perception to spatial quantization of facial images. *Psychology, 1*(3), 199–208.

Harmon, L. D. (1971). Some aspects of recognition of human faces. In O. J. Grüsser, & R. Klinke (Eds.), *Pattern recognition in biological and technical systems*. Heidelberg: Springer.

Harmon, L. D. (1973). The recognition of faces. *Scientific American, 229*, 71–82.

Harmon, L. D., & Julesz, B. (1973). Masking in visual recognition: Effects of two-dimensional filtered noise. *Science, 180*, 1194–1197.

Harris, C. S. (Ed.), (1980). *Visual coding and adaptability* Hillsdale, NJ: LEA.

Hegdé, J. (2008). Time course of visual perception: Coarse-to-fine processing and beyond. *Progress in Neurobiology, 84*, 405–439.

Heinze, H.-J., & Münte, T. F. (1993). Electrophysiological correlates of hierarchical stimulus processing: Dissociation between onset and later stages of global and local target processing. *Neuropsychologia, 31*, 841–852.

Henderson, Z., Bruce, V., & Burton, A. M. (2001). Matching the faces of robbers captured on video. *Applied Cognitive Psychology, 15,* 445–464.

Hess, R. E. (2004). Spatial scale in visual processing. In L. M. Chalupa, & J. S. Werner (Eds.), *The visual neurosciences* (pp. 1043–1059). Cambridge, MA: MIT Press.

Hoar, S., & Linnell, K. J. (2013). Cognitive load eliminates the global perceptual bias for unlimited exposure durations. *Attention, Perception, & Psychophysics, 75,* 210–215.

Hoffman, J. E. (1980). Interaction between global and local levels of a form. *Journal of Experimental Psychology: Human Perception and Performance, 6,* 222–234.

Holmes, A., Winston, J. S., & Eimer, M. (2005). The role of spatial frequency information for ERP components sensitive to faces and emotional facial expression. *Cognitive Brain Research, 25,* 508–520.

Hsiao, F.-J., Hsieh, J.-C., Lin, Y.-Y., & Chang, Y. (2005). The effects of spatial frequencies on cortical processing revealed by magnetoencephalography. *Neuroscience Letters, 380,* 54–59.

Hübner, R. (1997). The effect of spatial frequency on global precedence and hemispheric differences. *Perception & Psychophysics, 59,* 187–201.

Hughes, H. C., Fendrich, R., & Reuter-Lorenz, P. A. (1990). Global versus local processing in the absence of low spatial frequencies. *Journal of Cognitive Neuroscience, 2,* 272–282.

Hughes, H. C., Nozawa, G., & Kitterle, F. (1996). Global precedence, spatial frequency channels, and the statistics of natural images. *Journal of Cognitive Neuroscience, 8,* 197–230.

Jacques, C., Schiltz, C., & Goffaux, V. (2014). Face perception is tuned to horizontal orientation in the N170 time window. *Journal of Vision, 14*(2), 1–18, 5.

Joubert, O. R., Rousselet, G. A., Fabre-Thorpe, M., & Fize, D. (2009). Rapid visual categorization of natural scene contexts with equalized amplitude spectrum and increasing phase noise. *Journal of Vision, 9*(1), 2 . 1–16, http://journalofvision.org/9/1/2/, http://dx.doi.org/10.1167/9.1.2

Judd, T., Durand, F., & Torralba, A. (2011). Fixations on low-resolution images. *Journal of Vision, 11,* 1–20. Available from http://dx.doi.org/10.1167/11.4.14.

Kauffmann, L., Chauvin, A., Pichat, C., & Peyrin, C. (2015). Effective connectivity in the neural network underlying coarse-to-fine categorization of visual scenes. A dynamic causal modeling study. *Brain and Cognition, 99,* 46–56.

Kimchi, R. (1992). Primacy of wholistic processing and global/local paradigm. *Psychological Bulletin, 112,* 24–38.

Kimchi, R. (1998). Uniform connectedness and grouping in the perceptual organization of hierarchical patterns. *Journal of Experimental Psychology: Human Perception and Performance, 24,* 1105–1118.

Kimchi, R., & Palmer, S.-E. (1982). Form and texture in hierarchically constructed patterns. *Journal of Experimental Psychology: Human Perception and Performance, 8,* 521–535.

Kinchla, R. A., & Wolfe, J. M. (1979). The order of visual processing: "top-down", "bottom-up" or "middle-out". *Perception and Psychophysics, 25,* 225–231.

Knoche, H., McCarthy, J., & Sasse, M.A. (2005). Can small be beautiful? Assessing image resolution requirements for mobile TV. In *ACM multimedia ACM.*

Knoche, H., McCarthy, J. D., & Sasse, M. A. (2008). How low can you go? The effect of low resolutions on shot types in mobile TV. *Multimedia Tools and Applications, 36,* 145–166.

Koivisto, M., & Revonsuo, A. (2004). Preconscious analysis of global structure: Evidence from masked priming. *Visual Cognition, 11,* 105–127.

Lakshminarayanan, V., Bhatia, S. K., Samal, A., & Welland, G. V. (1997). *Reaction times for recognition of degraded facial images. Basic and clinical applications of vision science* (pp. 287–293). Netherlands: Springer.

Lamb, M. R., London, B., Pond, H. M., & Whitt, K. A. (1998). Automatic and controlled processes in the analysis of hierarchical structure. *Psychological Science, 9,* 14–19.

Lamb, M. R., Pond, H. M., & Zahir, G. (2000). Contributions of automatic and controlled processes to the analysis of hierarchical structure. *Journal of Experimental Psychology: Human Perception and Performance, 26,* 1234–1245.

Lamb, M. R., & Yund, E. W. (1996a). Spatial frequency and interference between global and local levels of structure. *Visual Cognition, 3,* 193–219.

Lamb, M. R., & Yund, E. W. (1996b). Spatial frequency and attention: Effects of level-, target-, and location-repetition on the processing of global and local forms. *Perception and Psychophysics, 58,* 363–373.

Lamb, M. R., Yund, E. W., & Pond, H. M. (1999). Is attentional selectivity to different levels of hierarchical structure based on spatial frequency? *Journal of Experimental Psychology: General, 128,* 88–94.

Lander, K., Bruce, V., & Hill, H. (2001). Evaluating the effectiveness of pixelation and blurring on masking the identity of familiar faces. *Applied Cognitive Psychology, 15,* 101–116.

Large, M.-E., & McMullen, P. A. (2006). Hierarchical attention in discriminating objects at different levels of specificity. *Perception & Psychophysics, 68,* 845–860.

Leder, H. (1996). Line drawings of faces reduce configural processing. *Perception, 25,* 355–366.

Leder, H., & Bruce, V. (2000). When inverted faces are recognized: The role of configural information in face recognition. *Quarterly Journal of Experimental Psychology, 53A,* 513–536.

Lee, W. J., Wilkinson, C., Memon, A., & Houston, K. (2009). Matching unfamiliar faces from poor quality closed-circuit television (CCTV) footage. *AXIS, 1,* 19–28.

Leeuwenberg, E., Mens, L., & Calis, G. (1985). Knowledge within perception: Masking caused by incompatible interpretation. *Acta Psychologica, 55,* 91–102.

Levelt, W. J. M. (1968). *On binocular rivalry.* The Hague: Mouton.

Liu, C. H., Collin, C. A., Rainville, S. J. M., & Chaudhuri, A. (2000). The effects of spatial frequency overlap on face recognition. *Journal of Experimental Psychology: Human Perception and Performance, 26,* 956–979.

Liu, J., Harris, A., & Kanwisher, N. (2002). Stages of processing in face perception: An MEG study. *Nature Neuroscience, 5,* 910–916.

Loftus, G. R., & Harley, E. M. (2004). How different spatial-frequency components contribute to visual information acquisition. *Journal of Experimental Psychology: Human Perception and Performance, 30,* 104–118.

Love, B. C., Rouder, J. N., & Wisniewski, E. J. (1999). A structural account of global and local processing. *Cognitive Psychology, 38,* 291–316.

Lu, Z.-L., & Sperling, G. (1996). The Lincoln picture non-problem. *Investigative Ophtalmology and Visual Science, ARVO Supplement, 37*(3), 732.

MacDonald, J., Andersen, S., & Bachmann, T. (2000). Hearing by eye: How much spatial degradation can be tolerated? *Perception, 29,* 1155–1168.

Madisson, K. (2010). *Nägude äratundmine.* Uurimistöö. Tartu Ülikool, Õigusteaduskond, Avaliku õiguse instituut.

Marendaz, C. (1985). Précedence globale et dependance du champ: Des routines visuelles? *Cahiers de Psychologie Cognitive, 5,* 727–745.

Martens, U., & Hübner, R. (2013). Functional hemispheric asymmetries of global/local processing mirrored by the steady-state visual evoked potential. *Brain and Cognition, 81*, 161–166.

Martin, M. (1979). Local and global processing: The role of sparsity. *Memory and Cognition, 7*, 476–484.

Maurer, D., Le Grand, R., & Mondloch, C. J. (2002). The many faces of configural processing. *Trends in Cognitive Sciences, 6*, 255–260.

Maurer, D., O'Craven, K. M., Le Grand, R., Mondloch, C. J., Springer, M. V., Lewis, T. L., et al. (2007). Neural correlates of processing facial identity based on features versus their spacing. *Neuropsychologia, 45*, 1438–1451.

McCarthy, S. (2005). The art portrait, the pixel and the gene: Micro construction of macro representation. *Convergence, 11*, 60–71.

McGurk, H., & MacDonald, J. (1976). Hearing lips and seeing voices. *Nature, 264*, 746–748.

McKone, E., Davies, A. A., Darke, H., Crookes, K., Wickramariyarante, T., Zappia, S., et al. (2013). Importance of the inverted control in measuring holistic face processing with the composite effect and part-whole effect. *Frontiers in Psychology, 4*, 33. Available from http://dx.doi.org/10.3389/fpsyg.2013.00033.

Meinhardt-Injac, B., Persike, M., & Meinhardt, G. (2010). The time course of face matching by internal and external features: Effects of context and inversion. *Vision Research, 50*, 1598–1611.

Mermillod, M., Guyader, N., & Chauvin, A. (2005). The coarse-to-fine hypothesis revisited: Evidence from neuro-computational modeling. *Brain and Cognition, 57*, 151–157.

Michaels, C. F., & Turvey, M. T. (1979). Central sources of visual masking: Indexing structures supporting seeing at a single, brief glance. *Psychological Research, 41*, 1–61.

Miellet, S., Caldara, R., & Schyns, P. G. (2011). Local Jekyll and global Hyde: The dual identity of face identification. *Psychological Science, 22*, 1518–1526. Available from http://dx.doi.org/10.1177/0956797611424290.

Miller, J. (1981). Global precedence in attention and decision. *Journal of Experimental Psychology: Human Perception and Performance, 7*, 1161–1174.

Morgan, M. J., & Watt, R. J. (1997). The combinations of filters in early spatial vision: A retrospective analysis of the MIRAGE model. *Perception, 26*, 1073–1088.

Morrison, D. J., & Schyns, P. G. (2001). Usage of spatial scales for the categorization of faces, objects and scenes. *Psychonomic Bulletin and Review, 8*, 454–469.

Morrone, C., & Burr, D. C. (1997). Capture and transparency in coarse quantized images. *Vision Research, 37*, 2609–2629.

Morrone, M. C., & Burr, D. C. (1994). Visual capture and transparency in blocked images. *Perception, 23S*, 20b.

Morrone, M. C., Burr, D. C., & Ross, J. (1983). Added noise restores recognizability of coarse quantised images. *Nature, 305*, 226–228.

Mu, T., & Li, S. (2013). The neural signature of spatial frequency-based information integration in scene perception. *Experimental Brain Research, 227*, 367–377.

Munhall, K. G., Kroos, C., Jozan, G., & Vatikiotis-Bateson, E. (2004). Spatial frequency requirements for audiovisual speech perception. *Perception & Psychophysics, 66*, 574–583.

Musel, B., Kauffmann, L., Ramanoël, S., Giavarini, C., Guyader, N., Chauvin, A., et al. (2014). Coarse-to-fine categorization of visual scenes in scene-selective cortex. *Journal of Cognitive Neuroscicence, 26*(10), 2287–2297.

Nandakumar, C., & Malik, J. (2009). Understanding rapid category detection via multiply degraded images. *Journal of Vision, 9*(6). 19, 1–8, http://journalofvision.org/9/6/19/, doi:10.1167/9.6.19

Näsänen, R. (1999). Spatial frequency bandwidth used in the recognition of facial images. *Vision Research, 39*, 3824–3833.

Navon, D. (1977). Forest before trees: The precedence of global features in visual perception. *Cognitive Psychology, 9*, 353–383.

Navon, D. (1981a). The forest revisited: More on global precedence. *Psychological Research, 43*, 1–32.

Navon, D. (1981b). Do attention and decision follow perception? Comment on Miller. *Journal of Experimental Psychology: Human Perception and Performance, 7*, 1175–1182.

Navon, D. (1983). How many trees does it take to make a forest? *Perception, 12*, 239–254.

Navon, D., & Norman, J. (1983). Does global precedence really depend on visual angle? *Journal of Experimental Psychology: Human Perception and Performance, 9*, 955–965.

Neri, P. (2011). Coarse to fine dynamics of monocular and binocular processing in human pattern vision. *Proceedings of the National Academy of Sciences USA, 108*, 10726–10731.

Noll, A.M. (2013). Early digital computer art & animation at Bell Telephone Laboratories, Inc. posted on Internet, June 17, 2013.

Nurmoja, M., Eamets, T., Härma, H.-L., & Bachmann, T. (2012). Dependence of the appearance-based perception of criminality, suggestibility, and trustworthiness on the level of pixelation of facial images. *Perceptual & Motor Skills, 115*, 465–480.

Nuutinen, M., Halonen, R., Leisti, T., & Oittinen, P. (2010). Reduced-reference quality metrics for measuring the image quality of digitally printed natural imagesIn S. P. Farnand, & F. Gaykema (Eds.), *Image quality and system performance VII*, Proceedings of SPIE-IS&T electronic imaging (SPIE Vol. 7529, p. 752901).

Oliva, A., & Schyns, P. G. (1997). Coarse blobs or fine edges? Evidence that information diagnosticity changes the perception of complex visual stimuli. *Cognitive Psychology, 34*, 72–107.

Olshausen, B. A. (2004). Principles of image representation in visual cortex. In L. M. Chalupa, & J. S. Werner (Eds.), *The visual neurosciences* (pp. 1603–1615). Cambridge, MA: MIT Press.

Otsuka, Y., Ichikawa, H., Kanazawa, S., Yamaguchi, M. K., & Spehar, B. (2014). Temporal dynamics of spatial frequency processing in infants. *Journal of Experimental Psychology: Human Perception and Performance, 40*(3), 995–1008.

Oudaya Coumar, S., Rajesh, P., Balaji, S. & Sadanandam, S. (2013). Image restoration using filters and image quality assessment using reduced reference metrics. In: *2013 International conference on circuits, controls and communications (CCUB), Bengaluru, 27–28 December* (pp. 1–5). doi:10.1109/CCUBE.2013.6718542.

Palmeri, T. J., & Gauthier, I. (2004). Visual object understanding. *Nature Reviews Neuroscience, 5*, 291–303.

Paquet, L. (1999). Global dominance outside the focus of attention. *Quarterly Journal of Experimental Psychology, 52A*, 465–485.

Paquet, L., & Merikle, P. (1988). Global precedence in attended and nonattended objects. *Journal of Experimental Psychology: Human Perception and Performance, 14*, 89–100.

Paquet, L., & Merikle, P. M. (1984). Global precedence: The effect of exposure duration. *Canadian Journal of Psychology, 38*, 45–53.

Parker, D. M., & Costen, N. P. (1999). One extreme or the other or perhaps the golden mean? Issues of spatial resolution in face processing. *Current Psychology, 18*, 118–127.

Parker, D. M., Lishman, J. R., & Hughes, J. (1992). Temporal integration of spatially filtered visual images. *Perception, 21*, 147–160.

Parker, D.M., Lishman, J.R., & Hughes, J. (1996a). Integration of spatial information in human vision is temporally anisotropic: Evidence from a spatio-temporal discrimination task. In: *Workshop on spatial scale interactions, Centre for vision and visual cognition, University of Durham, September 16−18.* p. 15.

Parker, D. M., Lishman, J. R., & Hughes, J. (1996b). Role of coarse and fine spatial information in face and object processing. *Journal of Experimental Psychology: Human Perception and Performance, 22,* 1448−1466.

Parker, D. M., Lishman, J. R., & Hughes, J. (1997). Evidence for the view that temporospatial integration in vision is temporally anisotropic. *Perception, 26,* 1169−1180.

Pashler, H. (1998). *The psychology of attention.* Cambridge, MA: Bradford/MIT Press.

Pelli, D. G. (1999). Close encounters: An artist shows that size affects shape. *Science, 285*(5429), 844−846.

Pérez Fornos, A., Sommerhalder, J., Rappaz, B., Safran, A. B., & Pelizzone, M. (2005). Simulation of artificial vision, III: Do the spatial or temporal characteristics of stimulus pixelization really matter? *Investigative Ophtalmology and Visual Science, 46,* 3906−3912.

Piepers, D. W., & Robbins, R. A. (2012). A review and clarification of the terms "holistic", "configural", and "relational" in the face perception literature. *Frontiers in Psychology, 3,* 559. Available from http://dx.doi.org/10.3389/fpsyg.2012.00559.

Pratt, W. (2014). *Introduction to digital image processing.* Boca Raton, FL: CRC Press.

Rakover, S. S. (2002). Featural vs. configurational information in faces: A conceptual and empirical analysis. *British Journal of Psychology, 93,* 1−30.

Ramachandran, V. S., & Rogers-Ramachandran, D. (2006). Cracking the Da Vinci code. *Scientific American, Junel July,* 14−16.

Ravin, J. G., & Odell, P. M. (2008). Pixels and painting. Chuck Close and the fragmented image. *Archives in Ophtalmology, 126,* 1148−1151.

Richler, J. J., Palmeri, T. J., & Gauthier, I. (2012). Meanings, mechanisms, and measures of holistic processing. *Frontiers in Psychology, 3,* 553. Available from http://dx.doi.org/10.3389/fpsyg.2012.00553.

Riddernikhof, K. R., & van der Molen, M. W. (1995). When global information and local information collide: A brain potential analysis of the locus of interference effects. *Biological Psychology, 41,* 29−53.

Rijpkema, M., van Aalderen, S., Schwarzbach, J., & Verstraten, F. A. J. (2007). Beyond the forest and the trees: Local and global interference in hierarchical visual stimuli containing three levels. *Perception, 36,* 1115−1122.

Roberts, D. J., Woollams, A. M., Kim, E., Beeson, P. M., Rapcsak, S. Z., & Lambon Ralph, M. A. (2013). Efficient visual object and word recognition relies on high spatial frequency coding in the left posterior fusiform gyrus: Evidence from a case-series of patients with ventral occipito-temporal cortex damage. *Cerebral Cortex, 23,* 2568−2580.

Robertson, L. C., Egly, R., Lamb, M. R., & Kerth, L. (1993). Spatial attention and cuing to global and local levels of hierarchical structure. *Journal of Experimental Psychology: Human Perception and Performance, 19,* 471−487.

Robson, J. G. (1980). Neural images: The physiological basis of spatial vision. In S. C. Harris (Ed.), *Visual coding and adaptability* (pp. 177−214). Hillsdale, NJ: LEA.

Romei, V., Driver, J., Schyns, P. G., & Thut, G. (2011). Rhythmic TMS over parietal cortex links distinct brain frequencies to global versus local visual processing. *Current Biology, 21* 334−337.

Ross, D. A., & Gauthier, I. (2015). Holistic processing in the composite task depends on face size. *Visual Cognition*, *23*(5), 533−545.

Rotshtein, P., Vuilleumier, P., Winston, J., Driver, J., & Dolan, R. (2007). Distinct and convergent visual processing of high and low spatial frequency informatikon in faces. *Cerebral Cortex*, *17*, 2713−2724.

Ruiz-Soler, M., & Beltran, F. S. (2006). Face perception: An integrative review of the role of spatial frequencies. *Psychological Research*, *70*, 273−292.

Sanocki, T. (1993). Time course of object identification: Evidence for a global-to-local contingency. *Journal of Experimental Psychology: Human Perception and Performance*, *19*, 878−898.

Schiano D.J., Ehrlich S.M., Sheridan K., 2004 Categorical imperative NOT: Facial affect is perceived continuously. In *Conference on Human factors in computing systems: Proceedings of the SIGCHI conference on Human factors in computing systems, Vienna, Austria* (pp. 49−56). doi. acm.org/10.1145/985692.985699.

Schiltz, C., & Rossion, B. (2006). Faces are represented holistically in the human occipito-temporal cortex. *NeuroImage*, *32*, 1385−1394.

Schwaninger, A., Ryf, S., & Hofer, F. (2003). Configural information is processed differently in perception and recognition of faces. *Vision Research*, *43*, 1501−1505.

Schwartz, L. F. (1998). Computer-aided illusions: Ambiguity, perspective and motion. *The Visual Computer*, *14*, 52−68.

Schyns, P. G. (1997). Categories and percepts: A bi-directional framework for categorization. *Trends in Cognitive Sciences*, *1*, 183−189.

Schyns, P. G., Goldstone, R. L., & Thibaut, J. P. (1998). The development of features in object concepts. *Behavioral and Brain Sciences*, *21*, 17−41.

Schyns, P. G., & Oliva, A. (1994). From blobs to boundary edges: Evidence for time- and spatial-scale-dependent scene recognition. *Psychological Science*, *5*, 195−200.

Schyns, P. G., & Oliva, A. (1999). Dr. Angry and Mr. Smile: When categorization flexibly modifies the perception of faces in rapid visual presentations. *Cognition*, *69*, 243−265.

Sergent, J. (1984). An investigation into component and configural processes underlying face perception. *British Journal of Psychology*, *75*, 221−242.

Sergent, J. (1986). Microgenesis of face perception. In H. Ellis, M. Jeeves, F. Newcombe, & A. Young (Eds.), *Aspects of face processing* (pp. 17−33). Dordrecht: Martinus Nijhoff.

Shahangian, K., & Oruc, I. (2014). Looking at a blurry old family photo? Zoom out!. *Perception*, *43*, 90−98.

Sheikh, H. R., & Bovik, A. C. (2006). Image information and visual quality. *IEEE Transactions in Image Processing*, *15*, 430−444.

Sheikh, H. R., Bovik, A. C., & De Veciana, G. (2005). An information fidelity criterion for image quality assessment using natural scene statistics. *IEEE Transactions in Image Processing*, *14*, 2117−2118.

Shulman, G. L., & Wilson, J. (1987a). Spatial frequency and selective attention to local and global information. *Perception*, *16*, 89−101.

Silpa, K., & Aruna Mastani, S. (2012). Comparison of image quality metrics. *International Journal of Engineering Research & Technology (IJERT)*, *1*(4), 1−5.

Sinha, P. (2002a). Recognizing complex patterns. *Nature Neuroscience Supplement*, *5*, 1093−1097.

Sinha, P. (2002b). Identifying perceptually significant features for recognizing faces. *Proceedings of SPIE Electronic Imaging Symposium*, *4662*, 12−21.

Sinha, P., Balas, B. J., Ostrovsky, Y., & Russell, R. (2006). Face recognition by humans. In W. Zhao, & R. Chellappa (Eds.), *Face processing: Advanced modeling methods* (pp. 257–291). San Diego, CA: Academic Press/Elsevier.

Smilek, D., Rempel, M. I., & Enns, J. T. (2006). The illusion of clarity: Image segmentation and edge attribution without filling-in. *Visual Cognition, 14*, 1–36.

Smith, A. P. (1985). The effects of noise on the processing of global shape and local detail. *Psychological Research, 47*, 103–108.

Solomon, C. J., & Breckon, T. P. (2010). *Fundamentals of digital image processing: A practical approach with examples in Matlab.* Chichester, UK: Wiley-Blackwell.

Sommerhalder, J. (2007). How to restore reading with visual prostheses. In J. Tombrass-Tink, et al. (Eds.),*Visual prosthesis and ophthalmic devices: New hope in sight* (pp. 15–35). Totows, NJ: Humana Press.

Sowden, P. T., & Schyns, P. G. (2006). Channel surfing in the visual brain. *Trends in Cognitive Sciences, 10*, 538–545.

Sperling, G., & Hsu, A. (2004). Revisiting the Lincoln Picture Problem. *Journal of Vision, 4*(8), 53a.

Stoffer, T. H. (1993). The time course of attentional zooming: A comparison of voluntary and involuntary allocation of attention to the levels of compound stimuli. *Psychological Research/ Psychologische Forschung, 56*, 14–25.

Stoffer, T. H. (1994). Attention zooming and the global-dominance phenomenon: Effects of level-specific cueing and abrupt visual onset. *Psychological Research/Psychologische Forschung, 56*, 83–98.

Sugase, Y., Yamane, S., Ueno, S., & Kawano, K. (1999). Global and fine information coded by single neurons in the temporal visual cortex. *Nature, 400*, 869–873.

Tanaka, J. W., & Farah, M. J. (1993). Parts and wholes in face recognition. *Quarterly Journal of Experimental Psychology, 46*, 225–245.

Thomas, C., Kveraga, K., Huberle, E., Karnath, H.-O., & Bar, M. (2012). Enabling global processing in simultanagnosia by psychophysical biasing of visual pathways. *Brain, 135*, 1578–1585.

Thompson, R. W., Barnett, G. D., Humayun, M. S., & Dagnelie, G. (2003). Facial recognition using simulated prosthetic pixelized vision. *Investigative Opthalmology & Visual Science, 44*, 5035–5042.

Tieger, T., & Ganz, L. (1979). Recognition of faces in the presence of two-dimensional sinusoidal masks. *Perception and Psychophysics, 26*, 163–167.

Todorov, A., Said, C. P., & Verosky, S. C. (2011). Personality impressions from facial appearance. In A. J. Calder, G. Rhodes, M. H. Johnson, & J. V. Haxby (Eds.), *The Oxford handbook of face perception* (pp. 631–652). Oxford, UK: Oxford University Press.

Torralba, A. (2009). How many pixels make an image? *Visual Neuroscience, 26*, 123–131.

Turvey, M. T. (1973). On peripheral and central processes in vision: Inferences from an information-processing analysis of masking with patterned stimuli. *Psychological Review, 80*, 1–52.

Uttal, W. R. (1981). *A taxonomy of visual processes.* Hillsdale, NJ: Erlbaum.

Uttal, W. R. (1988). *On seeing forms.* Hillsdale, NJ: Erlbaum.

Uttal, W. R. (1994). *Pattern recognition.* Encyclopedia of human behavior (Vol. 3, pp. 421–429). San Diego, CA: Academic Press.

Uttal, W. R. (1998). What degraded images tell us about perceptual theories. In R. R. Hoffman, M. F. Sherrick, & J. S. Warm (Eds.), *Viewing psychology as a whole: The integrative science of William N. Dember* (pp. 19–43). Washington, DC: APA.

Uttal, W. R., Baruch, T., & Allen, L. (1995a). Combining image degradations in a recognition task. *Perception and Psychophysics, 57*, 682–691.

Uttal, W. R., Baruch, T., & Allen, L. (1995b). Dichoptic and physical information combination: A comparison. *Perception, 24*, 351–362.

Uttal, W. R., Baruch, T., & Allen, L. (1995c). The effect of combinations of image degradations in a discrimination task. *Perception and Psychophysics, 57*, 668–681.

Uttal, W. R., Baruch, T., & Allen, L. (1997). A parametric study of face recognition when image degradations are combined. *Spatial Vision, 11*, 179–204.

van der Schaaf, A., & van Hateren, J. H. (1996). Modelling the power spectra of natural images: Statistics and information. *Vision Research, 36*, 2759–2770.

Van Vleet, T. M., Hoang-duc, A. K., DeGutis, J., & Robertson, L. C. (2011). Modulation of non-spatial attention and the global/local processing bias. *Neuropsychologia, 49*, 352–359.

Vassilev, A., & Stomonyakov, V. (1987). The effect of grating spatial frequency on the early VEP-component CI. *Vision Research, 27*, 727–729.

Vitkovich, M., & Barber, R. (1996). Visible speech as a function of image quality: Effects of display parameters on lipreading ability. *Applied Cognitive Psychology, 10*, 121–140.

Volberg, G., & Hübner, R. (2004). On the role of response conflicts and stimulus position for hemispheric differences in global/local processing: An ERP study. *Neuropsychologia, 42*, 1805–1813.

Vuilleumier, P., Armony, J. L., Driver, J., & Dolan, R. J. (2003). Distinct spatial frequency sensitivities for processing faces and emotional expressions. *Nature Neuroscience, 6*, 624–631. Available from http://dx.doi.org/10.1038/nn1057.

Wajid, R., Mansoor, A. B., & Pedersen, M. (2014). A human perception based performance evaluation of image quality metrics. In G. Bebis, et al. (Eds.), *ISVC 2014, Part I, LNCS 8887* (pp. 303–312). Cham (ZG), Switzerland: Springer International Publishing AG.

Walker, P. (1978). Binocular rivalry: Central or peripheral selective processes. *Psychological Bulletin, 85*, 376–389.

Wallbott, H. G. (1991). The robustness of communication of emotion via facial expression: Emotion recognition from photographs with deteriorated pictorial quality. *European Journal of Social Psychology, 21*, 89–98.

Wallbott, H. G. (1992). Effects of distortion of spatial and temporal resolution of video stimuli on emotion attributions. *Journal of Nonverbal Behavior, 16*, 5–20.

Wang, Z., Bovik, A. C., Sheikh, H. R., & Simoncelli, E. P. (2004). Image quality assessment: From error visibility to structural similarity. *IEEE Transactions in Image Processing, 13*(April), 1–12.

Ward, L. M. (1982). Determinants of attention to local and global features of visual forms. *Journal of Experimental Psychology: Human Perception and Performance, 8*, 562–581.

Ward, L. M. (1983). On processing dominance: Comment on Pomerantz. *Journal of Experimental Psychology: General, 112*, 541–546.

Watson, A. B. (1986). Temporal sensitivity. In K. R. Boff, L. Kaufman, & J. P. Thomas (Eds.), *Handbook of perception and human performance. Vol. 1. Sensory processes and perception (6/1-6/ 43)*. New York, NY: Wiley.

Watson, T. L., & Robbins, R. A. (2014). The nature of holistic processing in face and object recognition: Current opinions. *Frontiers in Psychology, 5*, 3. Available from http://dx.doi.org/ 10.3389/fpsyg.2014.00003.

Watt, R. (1988). *Visual processing*. London: Erlbaum.

Weisstein, N. (1980). The joy of fourier analysis. In S. C. Harris (Ed.), *Visual coding and adaptability* (pp. 365–380). Hillsdale, NJ: LEA.

Werner, H. (1940). Studies on contours: Strobostereoscopic phenomena. *The American Journal of Psychology, 53,* 418–422.

Werner, J. S., & Chalupa, L. M. (2013). *The new visual neurosciences.* Cambridge, MA: MIT Press.

Westheimer, G. (2012). Spatial and spatial-frequency analysis in visual optics. *Ophthalmic and Physiological Optics, 32,* 271–281. Available from http://dx.doi.org/10.1111/j.1475-1313.2012.00913.x.

White, M., & Li, J. (2006). Matching faces and expressions in pixelated and blurred photos. *American Journal of Psychology, 119,* 21–28.

Williams, D. R., & Hofer, H. (2004). Formation and acquisition of the retinal image. In L. M. Chalupa, & J. S. Werner (Eds.), *The visual neurosciences* (pp. 795–810). Cambridge, MA: MIT Press.

Yang, M.-H., Kriegman, D. J., & Ahuja, N. (2002). Detecting faces in images: A survey. *IEEE Transactions on Pattern Analysis and Machine Intelligence, 2002*(24), 34–58.

Young, A. M., Hellawell, D., & Hay, D. C. (1987). Configural information in face perception. *Perception, 10,* 747–759.

Zebrowitz, L. A. (2011). Ecological and social approaches to face perception. In A. J. Calder, G. Rhodes, M. H. Johnson, & J. V. Haxby (Eds.), *The Oxford handbook of face perception* (pp. 31–50). Oxford, UK: Oxford University Press.

Zhang, L., Zhang, L., Mou, X., & Zhang, D. (2011). FSIM: A feature similarity index for image quality assessment. *IEEE Transactions in Image Processing, 20*(8), 2378–2386.

Zhao, M., Cheung, S.-h, Wong, A. C.-N., Rhodes, G., Chan, E. K. S., Chan, W. W. L., et al. (2014). Processing of configural and componential information in face-selective cortical areas. *Cognitive Neuroscience, 5*(3–4), 160–167.

Zhao, W., & Chellappa, R. (Eds.), (2006). *Face processing: Advanced modeling methods.* San Diego, CA: Academic Press/Elsevier.

Printed in the United States
By Bookmasters